T203 BLOCK 1: SCIENCE FOR MATERIALS
Units 5 and 6: Materials and chemistry

Unit 5   The Periodic Table and chemical bonding

Unit 6   Chemical reactions

The Open University

The Open University, Walton Hall, Milton Keynes, MK7 6AA

First published 1994. Reprinted 1997 (with corrections), 2001

Copyright © 1994 The Open University

All rights reserved. No part of this publication may be reproduced, stored in a retrieval system or transmitted in any form or by any means without written permission from the publisher or without a licence from the Copyright Licensing Agency Limited. Details of such licences (for reprographic reproduction) may be obtained from the Copyright Licensing Agency Ltd, 90 Tottenham Court Road, London W1P 0LP.

Edited, designed and typeset by The Open University.

Printed in the United Kingdom by the Alden Press, Oxford

This text forms part of an Open University Second Level Course. If you would like a copy of *Studying with the Open University*, please write to the Central Enquiry Service, PO Box 200, The Open University, Walton Hall, Milton Keynes, MK7 6YZ, United Kingdom. If you have not already enrolled on the Course and would like to buy this or other Open University material, please write to Open University Worldwide, The Berrill Building, Walton Hall, Milton Keynes, MK7 6AA (Telephone 01908 858785; Facsimile 01908 858787; E-mail address ouwenq@open.ac.uk).

ISBN 0 7492 6205 2

Edition 1.3

21769C/t203u5-u6i1.3

T203 Course Team

ACADEMICS

Dr Nicholas Braithwaite
Dr Lyndon Edwards
Dr Andrew Greasley
Dr Adrian Hopgood
Dr Peter Lewis
Carol Morris (Staff Tutor, West Midlands Region)
Professor Bill Plumbridge
Ken Reynolds
Graham Weaver
Dr George Weidmann (Course Team Chair)

CRITICAL READERS

Chris Conway (Shrewsbury School)
Dr Peter Lederer (DRA, Malvern)
Dr John Spears (University of Middlesex)
Dr Peter Strain-Clark (Faculty of Mathematics, OU)

EXTERNAL ASSESSOR

Professor Arthur Willoughby (University of Southampton)

PRODUCTION

Phil Ashby (Producer, BBC)
Cameron Balbirnie (Producer, BBC)
Pam Berry (Text Processing Services)
Gail Block (Producer, BBC)
Andy Harding (Course Manager)
Caryl Hunter-Brown (Liaison Librarian)
Roy Lawrance (Graphic Artist)
Mike Levers (Photographer)
Carol Russell (Editor)
Karen Shipp (Academic Computing Services)
Ian Spratley (Producer, BBC)
John Stratford (Producer)
Bob Walters (Producer, BBC)
Rob Williams (Designer)

SECRETARY

Tracy Bartlett

TECHNICAL STAFF

Richard Black

Unit 5: The Periodic Table and chemical bonding

Contents

5.1	Introduction	6
5.2	Atoms and elements	6
5.3	Building the Periodic Table	20
5.4	The Periodic Table and electronic structure of the elements	28
5.5	Chemical bonding and periodicity	41
	Objectives for Unit 5	58
	Answers to exercises	60
	Answers to self-assessment questions	62

SCIENCE FOR MATERIALS

5.1 Introduction

So far, you've seen how materials respond to mechanical forces and how they interact with thermal energy. What you haven't yet considered in any detail is what the materials themselves are made of, and how this affects their properties.

In this unit you will examine the basic building blocks of materials, namely atoms, starting with the abundances and sources of the different types of atom (i.e. the chemical elements). You will then consider how the elements can be arranged in a classification scheme, and how this can be interpreted in terms of their underlying structure. This leads into an examination of how and why atoms join together to form molecules and compounds, and the different ways in which these are represented in chemistry. Along the way, I've included thumbnail sketches of various elements as they arise in the text, to give you a feel for the range and diversity of their properties and uses, especially in technology. Don't be put off by the large tables of data. They're only for reference and context; not to be learnt.

5.2 Atoms and elements

You are probably familiar with the idea that atoms are the basic building blocks of all materials. But what are atoms?

> **Exercise 5.1** What is meant by an atom? What are its principal constituents, what are their main characteristics and how are they distributed in the atom?

Of course, atoms are very small, as the following exercise reminds you.

> **Exercise 5.2** Given that the diameters of atoms are about 10^{-10} m, approximately how many atoms would fit in a 1 millimetre cube if the atoms are assumed to be spheres?

The answer to this exercise is only very approximate, since it assumes that the atoms would fill all the available space. But of course there are always voids when spheres pack together. You can see this in ▼**Close packing of atoms**▲, which examines how atoms can be arranged in the most space-efficient way.

The next exercise recalls another basic definition.

> **Exercise 5.3** What is meant by a chemical element? What single quantity is used to distinguish one element from another?

UNIT 5 THE PERIODIC TABLE AND CHEMICAL BONDING

▼Close packing of atoms▲

How atoms in solids are arranged relative to one another is a very important factor in determining the properties of materials. A regular, repeating arrangement leads to crystalline solids, a random arrangement to glassy ones, and a combination of the two is found in many polymers and some ceramics. In the solid state, the atoms of most elements pack together regularly to form different types of crystal structure, and many of these are what's known as **close-packed**. In other words, the atoms pack together to fill the available space to the maximum possible extent. Exercise 5.2 suggested that atoms can be modelled by spheres, implicitly assuming that they were incompressible. On a flat plane, such spheres of equal radius pack together most efficiently when each sphere is surrounded by six nearest neighbours (numbered 1 to 6 in Figure 5.1) giving a hexagonal array.

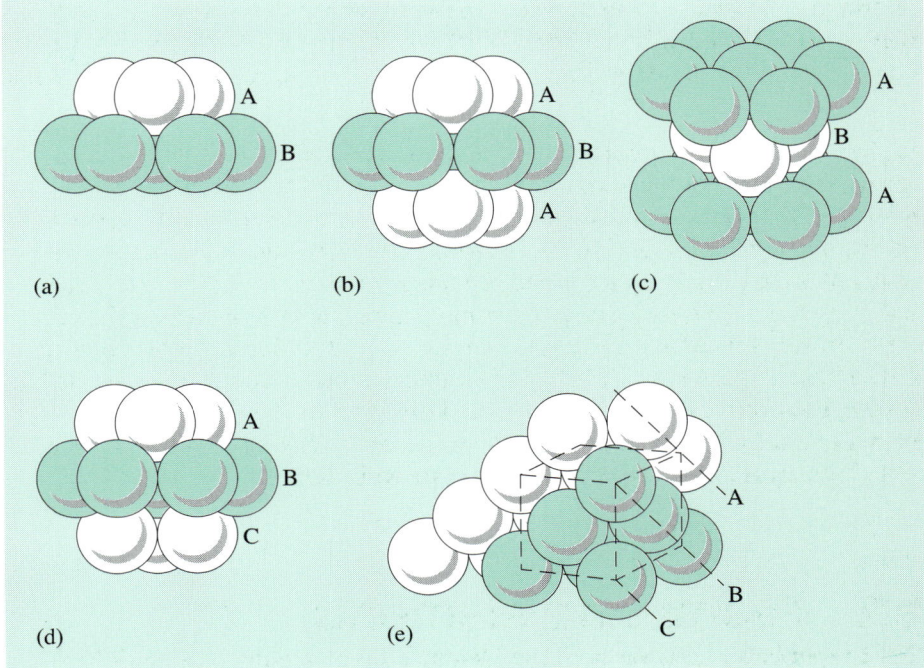

Figure 5.2 Two types of close packing

Figure 5.1 Two-dimensional close packing of spheres of equal size

Three-dimensional crystal structures are formed by stacking layers of such sheets. If one layer is labelled A and another layer labelled B, then A can fit into the hollows of B, as shown in Figure 5.2(a). Note that there are six hollows next to each atom, and that a second layer will fit into three of them, in alternate positions.

There are two possible basic stacking sequences, leading to two types of close-packed structure.

When a third layer is added so that it fits into the hollows on the other side of B with its atoms directly under those of the first layer (Figure 5.2(b)), it constitutes another A-layer. The sequence of layers is then ABABAB… and the resulting structure, shown in Figure 5.2 (c), is called **hexagonal close-packed** or **HCP**. Each atom is surrounded by twelve similar atoms; 6 in the same plane, 3 in the plane above and 3 in the plane below. The number of atoms immediately surrounding any given atom in a crystal structure is known as its **coordination number**, which in this case equals twelve.

In the second way of stacking the layers, the third layer can be located in the alternate set of hollows in B, so that its position (C) is different to either of the preceding layers (Figure 5.2(d)). The resulting sequence of layers is ABCABCABC…, which gives the **cubic close-packed** (or **CCP**) structure (Figure 5.2(e)).

You'll meet other types of crystal structure later in the course.

Where are the elements found and how much is there of them? On Earth, there are three sources of the elements: the Earth's crust, known as the lithosphere (from the Greek *lithos*, stone), which is about 35 km thick under continental land masses, and about 5 km thick under the oceans; the oceans themselves, together with other expanses of water, collectively known as the hydrosphere; and the atmosphere of gas surrounding it all. Table 5.1 shows the estimated quantities of the most abundant elements present in each of these spheres.

Table 5.1 The Earth's most abundant elements

Element	Symbol	Mass in the lithosphere / kg	Mass in the hydrosphere / kg	Mass in the atmosphere / kg	Total / % by mass
oxygen	O	1.1×10^{22}	1.5×10^{21}	1.2×10^{18}	49.1
silicon	Si	6.7×10^{21}	$<6.8 \times 10^{15}$		25.9
aluminium	Al	1.9×10^{21}	$<3.4 \times 10^{15}$		7.6
iron	Fe	1.2×10^{21}	$<3.4 \times 10^{13}$		4.7
calcium	Ca	8.6×10^{20}	6.8×10^{17}		3.4
sodium	Na	6.7×10^{20}	1.8×10^{19}		2.6
potassium	K	6.2×10^{20}	6.5×10^{17}		2.4
magnesium	Mg	5.0×10^{20}	2.2×10^{18}		2.0
titanium	Ti	9.6×10^{19}			0.4
hydrogen,	H	2.4×10^{19}	1.8×10^{20}	$\approx 5.1 \times 10^{9}$	0.8
phosphorus	P	2.4×10^{19}	$<3.4 \times 10^{14}$		
manganese	Mn	2.4×10^{19}	$<1.7 \times 10^{13}$		
chlorine	Cl	7.2×10^{18}	3.2×10^{19}		0.12
argon	Ar			5.1×10^{14}	
carbon	C	7.2×10^{18}	$<5.1 \times 10^{15}$	6.1×10^{14}	<.0.03
nitrogen	N	1.2×10^{18}	$<1.7 \times 10^{15}$	3.9×10^{18}	0.02
total		2.4×10^{22}	1.7×10^{21}	5.1×10^{18}	

The lithosphere is about 5000 times more massive than the atmosphere and about ten times more massive than the hydrosphere. If you recall from Unit 2 that the mass of the Earth is $m_E = 6 \times 10^{24}$ kg, you can see that Table 5.1 represents just over 0.4 % of the total — the rest is inaccessible to us. You should also bear in mind that the vast majority of the elements are present in chemically combined form. Free elements are a rarity.

You can see that ▼Oxygen▲ is far and away the most abundant element, comprising nearly half of the total. Although it is present as a free element in the atmosphere (of which it's about one-fifth), this mass is tiny compared to that of the oxygen chemically combined in the other two spheres — in the hydrosphere with ▼Hydrogen▲ to form water and in the lithosphere mainly with ▼Silicon▲ (the next most abundant element) to form silicates. The major constituent of the atmosphere is ▼Nitrogen▲ (80 % by volume), in which it too exists as a free element, but its overall abundance of 0.02 % is low. As you might expect, hydrogen and oxygen are the two most abundant elements in the hydrosphere (you should be familiar with the formula H_2O for water). The next two in abundance are ▼Sodium▲ and ▼Chlorine▲, reflecting the salinity of seawater.

▼Carbon▲ is one of a small number of elements that can exist as a free element in distinctly different forms — the phenomenon known as

▼Oxygen▲

Oxygen (symbol O, atomic number 8, density at s.t.p. 1.43 kg m^{-3}) was discovered in 1774 by Joseph Priestley, a British chemist and Presbyterian minister. It's a colourless, odourless, tasteless, diatomic (O_2) gas, although both the liquid (T_b = 90.2 K) and the solid (T_m = 54.8 K) are pale blue. It is essential for the life of animals and plants (with the exception of a few types of bacteria) and for most forms of combustion. In technological applications, steelmaking uses the largest amount of oxygen; to enrich the air injected into blast furnaces in making iron and in further refining the iron to make steel. It's also used in gas welding (oxy-acetylene), in medicine, in the synthesis of many important feedstock chemicals (e.g. methanol, ethylene oxide, ammonia) and as a component of rocket fuels. You'll see later that it is chemically very reactive and can combine with most of the other elements.

▼Hydrogen▲

It's estimated that about 75% of the mass of the universe and 90% of its atoms are hydrogen (symbol H, atomic number 1, density at s.t.p. 0.0899 kg m^{-3}). This makes it easily the most abundant element. All the other elements are derived from hydrogen, as the products of nuclear fusion reactions in stars. It's a colourless, odourless gas at s.t.p., is normally diatomic (H_2), and has the lowest density of any element. At very high pressures ($\approx$ 1 million atmospheres, or 10^{11} Pa), it's predicted that hydrogen transforms to a metallic solid, though this hasn't yet been confirmed experimentally. Technologically, it's used in the production of ammonia as well as in many other chemical processes, in the extraction of metals from their ores, as a welding gas, as a component of rocket fuels and, despite its flammability, as the gas for balloons.

▼Silicon▲

Silicon (symbol Si, atomic number 14, density 2330 kg m^{-3}) is a non-metallic element which was first isolated in 1824. It is a greyish, crystalline solid at room temperature, although it can also be prepared in amorphous (i.e. non-crystalline) form. Technologically, silicon is one of the most important and widely used elements. Being a semiconductor, it is one of the mainstays of the electronics industry (e.g. silicon chips), its electronic properties usually being modified by controlled additions of other elements (dopants). Silicon is also an important additive in steelmaking. Its oxide, silica is found naturally as sand, quartz, flint, agate, opal, etc. and is the principal ingredient of virtually all types of glass. (Remember the distinction between silicon and silica. They're often confused, as also are silicones, which are polymers of silicon and oxygen.) Silica together with silicates (from clays) are used to make cement and concrete as well as a whole range of ceramic products, from house bricks to porcelain and from furnace linings to toilet bowls. Silicon carbide is an important industrial abrasive material.

▼Nitrogen▲

Nitrogen (symbol N, atomic number 7, density at s.t.p. 1.25 kg m^{-3}) is a colourless, odourless and relatively inert gas, which, like oxygen and hydrogen, is normally diatomic (N_2). Because of its inertness, it is used as a 'blanket' gas in the electronics and pharmaceutical industries, and in certain types of beer dispensing. In contrast, compounds of nitrogen are very active, and are important in explosives, fertilizers, foodstuffs, drugs and poisons. Major tonnages are used in the production of ammonia (NH_3). Liquid nitrogen (T_b = 77.4 K) is also unreactive and colourless. It is widely use as a cryogenic (i.e. low-temperature) refrigerant.

▼Sodium▲

Sodium (symbol Na, atomic number 11, density 971 kg m^{-3}), whose name comes from soda, derived in turn from the Latin *sodanum*, headache remedy, is a very reactive, soft, bright silvery metal. Because of its high reactivity, it must be handled with care. It is the basis of many chemical processes, and many of its compounds are widely used. In its own right, it is used in sodium vapour lamps, in high-temperature nuclear reactors as a liquid coolant (T_m = 371 K) and in the cores of exhaust valves in high-performance automobile engines. It is also a reactant in one type of fuel cell (used to produce electrical energy directly from the energy of a chemical reaction).

▼Chlorine▲

Chlorine (symbol Cl, atomic number 17, density at s.t.p. 3.21 kg m^{-3}) is a reactive, greenish-yellow gas, normally diatomic (Cl_2), which was first identified as an element in 1810. It's a respiratory irritant at low concentrations, and is toxic at higher concentrations (it was used as a poison gas in the First World War). It is widely used for purifying water, for bleaching and in the production of paper, plastics (e.g. PVC), textiles, paints, insecticides, antiseptics, medicines, solvents, etc.

▼Carbon▲

Carbon (symbol C, atomic number 6) has been known in elemental form since prehistory. It is unique in the number (well over a million) and variety of compounds it can form, usually in association with hydrogen, oxygen and nitrogen. Their study is the separate branch of chemistry called organic chemistry. Without the ability of carbon atoms to combine with one another to create the backbones of long chain compounds (namely, polymers) life would not exist. This is also the basis of virtually all plastics. The presence of small amounts of carbon (around 1% by weight) is what distinguishes steel from iron and helps give it its range of controllable properties. The three principal allotropes of carbon (see 'Allotropy') are diamond, graphite and amorphous carbon.

Diamond has the highest values of Young's modulus and thermal conductivity of any material. It is also extremely hard. Apart from their value as gemstones, their hardness makes both synthetic and natural diamonds very useful industrially for cutting tools and as abrasives.

Graphite (which is the 'lead' in pencils) is used as a lubricant, as electrodes, as the brushes in electric motors and as a neutron absorber in nuclear reactors. It is also the basis for carbon fibres, whose high strength and Young's modulus are utilized in high performance composite materials. Uniquely, graphite's mechanical properties improve with increasing temperature, so it is used as the material for dies and plungers for high temperature pressing of ceramic and metal powders.

As charcoal, carbon is used by artists and also as an absorber of gases. Carbon black is added to plastics to absorb ultraviolet radiation and so prevent degradation, and to rubbers to reinforce them.

In some people's eyes, the relatively recently discovered buckminsterfullerene (after the US architect and proponent of geodesic structures, Buckminster Fuller) also ranks as an allotrope of carbon. This is a very stable molecule of 60 carbon atoms arranged in hexagons and pentagons around the surface of a sphere, very like the panels on a football or the structure of a geodesic sphere (Figure 5.3). It is the most stable of a family of related structures which have not yet found any uses, though there's no shortage of ideas.

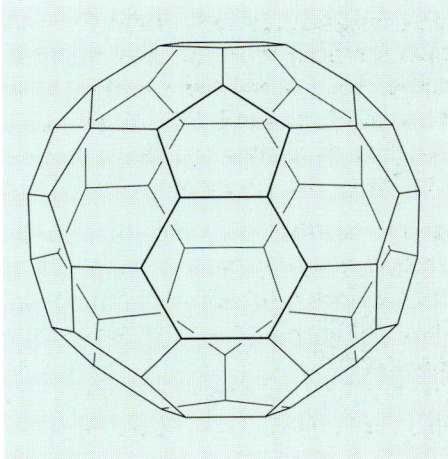

Figure 5.3 A 'bucky ball' or molecule of buckminsterfullerene, C_{60}

▼**Allotropy**▲. In view of carbon's importance to life, and its availability in oil, coal and natural gas, you might find its abundance surprisingly low. These fuels are all derived from living matter, and the total mass of living matter on Earth is only some 3.6×10^{14} kg. This is minute compared to any of the total masses shown in the bottom line of Table 5.1.

There are 92 naturally occurring elements, which I've listed alphabetically in Table 5.2. I've also shown their state at room temperature and atmospheric pressure, and their abundance as a percentage by mass in the Earth's crust (which is the dominant source for most of them). Of the 92, two (promethium and technetium) only occur naturally in stars, although they can be synthesized in nuclear reactors on Earth; ten, including the same two, are radioactive, meaning that they don't have any stable isotopes (I reminded you what isotopes are in the answer to Exercise 5.3); eleven are gases, only two are liquids and the remaining 79 are solids. Remarkably, 69 elements out of the 92 are metals, including one of the two liquids.

If you look at their relative abundances in the Earth's crust, you'll see that the numbers are consistent with those in Table 5.1 above. You should also be able to see something else. An element such as ▼**Gold**▲ is only present to the extent of 5×10^{-7}%, but, as you probably know, it can be found as a free, reasonably pure metal in the form of the nuggets and alluvial gold dust that prospectors seek. This must mean that it is not evenly dispersed throughout the lithosphere, but rather that there are local concentrations. The same is true of the other elements, and it is only this lumpy distribution, which reflects the chemical and physical reactions that they've undergone, that allows us selectively to extract the elements or element-bearing ores from the ground.

Also shown in Table 5.2 are the chemical symbols of each of the elements. These are explored further in ▼**Chemical symbols**▲.

Table 5.2 The naturally occurring elements

Element	Symbol	State*	Abundance in crust / % by mass
actinium ♦	Ac	s, m	—
aluminium	Al	s, m	8.1
antimony	Sb	s, m	1×10^{-4}
argon	Ar	g	4×10^{-5}
arsenic	As	s	5×10^{-4}
astatine ♦	At	s	—
barium	Ba	s, m	0.025
beryllium	Be	s, m	6×10^{-4}
bismuth	Bi	s, m	2×10^{-5}

* 'State' is at room temperature and atmospheric pressure: g, gas; l, liquid; s, solid; m, metal.

♦ All isotopes are radioactive

(continued opposite)

▼Allotropy▲

When an element can exist in two or more distinct solid forms, each type is known as an 'allotrope' and the phenomenon is called **allotropy**. Allotropes usually have quite different properties because the atoms are arranged in them in different ways. This usually means that they have different crystal structures, or no crystallinity at all. The difference in properties between allotropes can be so great as to suggest that they're different elements.

In the case of carbon, at least three forms are recognised: amorphous carbon, graphite and diamond. Each of three forms has a range of densities, depending on their source. Amorphous (or glassy) carbon, as its name suggests, is non-crystalline, and its density ranges from 1800 kg m^{-3} to 2100 kg m^{-3}. Graphite consists of sheet-like crystals, is very soft, black in colour and has a density in the range 1900 kg m^{-3} to 2300 kg m^{-3}. By contrast, diamond is an extremely hard, translucent crystalline material with a density range of 3150 kg m^{-3} to 3530 kg m^{-3}. That both graphite and diamond are forms of the same element was proved by Sir Humphrey Davy in 1820, who showed that each produced an identical compound, carbon dioxide, when burned in air.

Allotropy is normally found in non-metallic elements such as carbon, sulphur and phosphorus. However, some metals do show a related effect, by changing into different crystal structures at particular temperatures as they're heated or cooled (iron and tin provide good examples of this).

▼Gold▲

Gold (symbol Au, atomic number 79, density 19 300 kg m^{-3}) is a dense, inert, yellow metal. It's the most malleable and ductile of all the metals. (1 kg can be beaten out to about 1000 m^2. You might like to work out the thickness of the resulting gold leaf.) Its inertness has made it highly valued throughout history, and led to its use in coinage, jewellery and in decorative finishes (gilding or plating). More recent applications include dental work, infrared-reflecting coatings on spacecraft and the plating of electrical contacts to maintain their conductivity.

Table 5.2 The naturally occurring elements (continued)

Element	Symbol	State*	Abundance in crust / % by mass
boron	B	s	3×10^{-4}
bromine	Br	l	1.6×10^{-4}
cadmium	Cd	s, m	1.5×10^{-5}
caesium	Cs	s, m	7×0^{-4}
calcium	Ca	s, m	3.6
carbon	C	s	0.032
cerium	Ce	s, m	4.6×10^{-3}
chlorine	Cl	g	0.031
chromium	Cr	s, m	0.020
cobalt	Co	s, m	2.3×10^{-3}
copper	Cu	s, m	7.0×10^{-3}
dysprosium	Dy	s, m	4.5×10^{-4}
erbium	Er	s, m	2.5×10^{-4}
europium	Eu	s, m	1.1×10^{-4}
fluorine	F	g	0.030
francium ◆	Fr	s, m	—
gadolinium	Gd	s, m	6.4×10^{-4}
gallium	Ga	s, m	1.5×10^{-3}
germanium	Ge	s	7×10^{-4}
gold	Au	s, m	5×10^{-7}
hafnium	Hf	s, m	4.5×10^{-4}
helium	He	g	3×10^{-7}
holmium	Ho	s, m	1.2×10^{-4}
hydrogen	H	g	0.14
indium	In	s, m	1×10^{-5}
iodine	I	s	3×10^{-5}
iridium	Ir	s, m	1×10^{-7}
iron	Fe	s, m	5.0
krypton	Kr	g	—
lanthanum	La	s, m	1.8×10^{-3}
lead	Pb	s, m	1.6×10^{-3}
lithium	Li	s, m	6.5×10^{-3}
lutetium	Lu	s, m	8×10^{-5}
magnesium	Mg	s, m	2.09

* 'State' is at room temperature and atmospheric pressure: g, gas; l, liquid; s, solid; m, metal.

◆ All isotopes are radioactive

(continued overleaf)

Table 5.2 The naturally occurring elements (continued)

Element	Symbol	State*	Abundance in crust / % by mass
manganese	Mn	s, m	0.10
mercury	Hg	l, m	5×10^{-5}
molybdenum	Mo	s, m	1.5×10^{-3}
neodymium	Nd	s, m	2.4×10^{-3}
neon	Ne	g	—
nickel	Ni	s, m	8×10^{-3}
niobium	Nb	s, m	2.4×10^{-3}
nitrogen	N	g	4.6×10^{-3}
osmium	Os	s, m	1×10^{-7}
oxygen	O	g	46.6
palladium	Pd	s, m	1×10^{-6}
phosphorus	P	s	0.118
platinum	Pt	s, m	5×10^{-7}
polonium	Po	s, m	—
potassium	K	s, m	2.59
praseodymium	Pr	s, m	5.5×10^{-4}
(promethium) ♦	Pm	s, m	—
protactinium ♦	Pa	s, m	—
radium ♦	Ra	s, m	—
radon ♦	Rn	g	—
rhenium	Re	s, m	1×10^{-7}
rhodium	Rh	s, m	1×10^{-7}
rubidium	Rb	s, m	0.031
ruthenium	Ru	s, m	1×10^{-7}
samarium	Sm	s, m	6.5×10^{-4}
scandium	Sc	s, m	5×10^{-4}
selenium	Se	s	9×10^{-6}
silicon	Si	s	27.7
silver	Ag	s, m	1×10^{-5}
sodium	Na	s, m	2.83
strontium	Sr	s, m	0.03
sulphur	S	s	0.052
tantalum	Ta	s, m	2.1×10^{-4}
(technetium) ♦	Tc	s, m	—

* 'State' is at room temperature and atmospheric pressure: g, gas; l, liquid; s, solid; m, metal.

♦ All isotopes are radioactive

(continued opposite)

Table 5.2 The naturally occurring elements (continued)

Element	Symbol	State*	Abundance in crust / % by mass
tellurium	Te	s	2×10^{-7}
terbium	Tb	s, m	9×10^{-5}
thallium	Tl	s, m	6×10^{-5}
thorium ♦	Th	s, m	1.2×10^{-3}
thulium	Tm	s, m	2×10^{-5}
tin	Sn	s, m	4×10^{-3}
titanium	Ti	s, m	0.44
tungsten	W	s, m	6.9×10^{-3}
uranium ♦	U	s, m	4×10^{-4}
vanadium	V	s, m	0.015
xenon	Xe	g	—
ytterbium	Yb	s, m	2.7×10^{-4}
yttrium	Y	s, m	2.8×10^{-3}
zinc	Zn	s, m	0.0132
zirconium	Zr	s, m	0.022

* 'State' is at room temperature and atmospheric pressure: g, gas; l, liquid; s, solid; m, metal.
♦ All isotopes are radioactive

Now that you have a list of the elements and their symbols, you need to be able to order them in some way that shows a pattern in their behaviour. One basis for this comes from the answer to Exercise 5.3, the **atomic number**, which is given the symbol Z. Each increment in atomic number means that an extra proton has been added to the nucleus, in turn necessitating an extra electron for overall neutrality of electric charge. Another basis for ordering the elements might be **relative atomic mass** (r.a.m.), which is given the symbol A_r. Table 5.5 lists the elements, still in alphabetic order, along with their atomic numbers and relative atomic masses.

Exercise 5.4   What is meant by relative atomic mass?

You might think that relative atomic mass would provide just as good a basis for ranking the elements as atomic number. There is indeed a strong correlation between Z and A_r (Figure 5.4). But there are places (circled) where one element with a higher Z than another has isotopes with the same or lower A_r. Since the atomic number is unique to each element, it provides a more fundamental ranking. Table 5.6 shows the elements listed in order of increasing atomic number.

▼Chemical symbols▲

The symbols for the elements were shown in Table 5.1 and in the alphabetical listing of the elements in Table 5.2. The symbols consist of either one or two letters, the first one always written as a capital. Their great utility is that they provide a shorthand method for expressing the composition of chemical compounds, and of writing the reactants and products in the equations for chemical reactions. Table 5.3 lists the chemical symbols of the elements alphabetically, for your future reference.

Table 5.3 The naturally occurring elements by symbol

Symbol	Element	Atomic number, Z	Symbol	Element	Atomic number, Z	Symbol	Element	Atomic number, Z
Ac	actinium	89	He	helium	2	Ra	radium	88
Ag	silver	47	Hf	hafnium	72	Rb	rubidium	37
Al	aluminium	13	Hg	mercury	80	Re	rhenium	75
Ar	argon	18	Ho	holmium	67	Rh	rhodium	45
As	arsenic	33	I	iodine	53	Rn	radon	86
At	astatine	85	In	indium	49	Ru	ruthenium	44
Au	gold	79	Ir	iridium	77	S	sulphur	16
B	boron	5	K	potassium	19	Sb	antimony	51
Ba	barium	56	Kr	krypton	36	Sc	scandium	21
Be	beryllium	4	La	lanthanum	57	Se	selenium	34
Bi	bismuth	83	Li	lithium	3	Si	silicon	14
Br	bromine	35	Lu	lutetium	71	Sm	samarium	62
C	carbon	6	Mg	magnesium	12	Sn	tin	50
Ca	calcium	20	Mn	manganese	25	Sr	strontium	38
Cd	cadmium	48	Mo	molybdenum	42	Ta	tantalum	73
Ce	cerium	58	N	nitrogen	7	Tb	terbium	65
Cl	chlorine	17	Na	sodium	11	Tc	technetium	43
Co	cobalt,	27	Nb	niobium	41	Te	tellurium	52
Cr	chromium	24	Nd	neodymium	60	Th	thorium	90
Cs	caesium	55	Ne	neon	10	Ti	titanium	22
Cu	copper	29	Ni	nickel	28	Tl	thallium	81
Dy	dysprosium	66	O	oxygen	8	Tm	thulium	69
Er	erbium	68	Os	osmium	76	U	uranium	92
Eu	europium	63	P	phosphorus	15	V	vanadium	23
F	fluorine	9	Pa	protactinium	91	W	tungsten	74
Fe	iron	26	Pb	lead	82	Xe	xenon	54
Fr	francium	87	Pd	palladium	46	Y	yttrium	39
Ga	gallium	31	Pm	promethium	61	Yb	ytterbium	70
Gd	gadolinium	64	Po	polonium	84	Zn	zinc	30
Ge	germanium	32	Pr	praseodymium	59	Zr	zirconium	40
H	hydrogen	1	Pt	platinum	78			

You can see that, in the majority of cases, the first letter of the symbol is the initial letter of the element's name, and that where there's a second letter, it also occurs in the name. Largely for historical reasons, there are eleven elements whose symbols relate to alternative, usually Latin names. I've listed these in Table 5.4.

Table 5.4 Symbols related to other names

Symbol	Element	Origin
Ag	silver	Latin *argentum*
Au	gold	Latin *aurum*, 'shining dawn'
Cu	copper	Latin *cuprum*, 'from Cyprus'
Fe	iron	Latin *ferrum*
Hg	mercury	*hydragyrum*, 'liquid silver'
K	potassium	Latin *kalium*, 'alkali'
Na	sodium	Latin *natrium*
Pb	lead	Latin *plumbum*
Sb	antimony	Latin *stibium*, 'mark'
Sn	tin	Latin *stannum*
W	tungsten	wolfram, from the ore wolframite

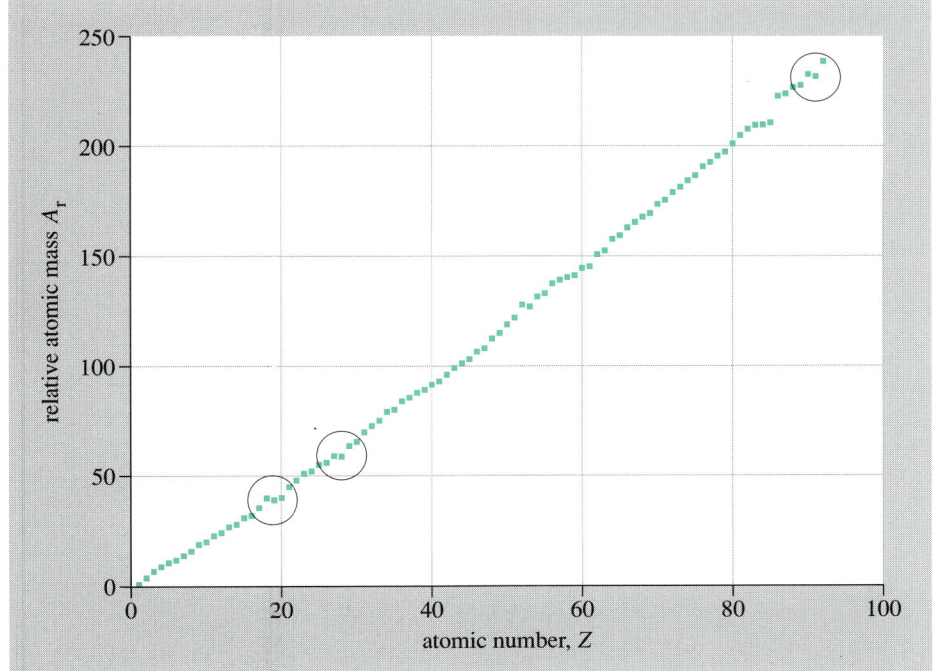

Figure 5.4 Relative atomic mass versus atomic number

Exercise 5.5 Assuming that it consists of just one isotope, how many protons and how many neutrons are there in the nucleus of an argon atom?

Table 5.5 Atomic numbers and relative atomic masses of the naturally occurring elements

Element	Symbol	Atomic number Z	Relative atomic mass A_r *
actinium ♦	Ac	89	227
aluminium	Al	13	26.7
antimony	Sb	51	122
argon	Ar	18	40.0
arsenic	As	33	74.9
astatine ♦	At	85	210
barium	Ba	56	137
beryllium	Be	4	9.01
bismuth	Bi	83	209
boron	B	5	10.8
bromine	Br	35	79.9
cadmium	Cd	48	112
caesium	Cs	55	133
calcium	Ca	20	40.1
carbon	C	6	12.0
cerium	Ce	58	140
chlorine	Cl	17	35.5
chromium	Cr	24	52.0
cobalt	Co	27	58.9
copper	Cu	29	63.6
dysprosium	Dy	66	163
erbium	Er	68	167
europium	Eu	63	152
fluorine	F	9	19.0
francium ♦	Fr	87	223
gadolinium	Gd	64	157
gallium	Ga	31	69.7
germanium	Ge	32	72.6
gold	Au	79	197
hafnium	Hf	72	178
helium	He	2	4.00
holmium	Ho	67	165
hydrogen	H	1	1.01
indium	In	49	115
iodine	I	53	127
iridium	Ir	77	192

* Atomic masses are rounded to three significant figures

♦ All isotopes are radioactive

(continued opposite)

Element	Symbol	Atomic number Z	Relative atomic mass A_r *
iron	Fe	26	55.9
krypton	Kr	36	83.8
lanthanum	La	57	139
lead	Pb	82	207
lithium	Li	3	6.94
lutetium	Lu	71	175
magnesium	Mg	12	24.3
manganese	Mn	25	54.9
mercury	Hg	80	201
molybdenum	Mo	42	95.9
neodymium	Nd	60	144
neon	Ne	10	20.2
nickel	Ni	28	58.7
niobium	Nb	41	92.9
nitrogen	N	7	14.0
osmium	Os	76	190
oxygen	O	8	16.0
palladium	Pd	46	106
phosphorus	P	15	31.0
platinum	Pt	78	195
polonium	Po	84	209
potassium	K	19	39.1
praseodymium	Pr	59	141
(promethium) ♦	Pm	61	145
protactinium ♦	Pa	91	231
radium ♦	Ra	88	226
radon ♦	Rn	86	222
rhenium	Re	75	186
rhodium	Rh	45	103
rubidium	Rb	37	85.5
ruthenium	Ru	44	101
samarium	Sm	62	150
scandium	Sc	21	45.0
selenium	Se	34	79.0
silicon	Si	14	28.1
silver	Ag	47	108
sodium	Na	11	23.0

* Atomic masses are rounded to three significant figures

♦ All isotopes are radioactive

(continued overleaf)

Table 5.5 Atomic numbers and relative atomic masses of the naturally occurring elements (continued)

Element	Symbol	Atomic number Z	Relative atomic mass A_r *
strontium	Sr	38	87.6
sulphur	S	16	32.1
tantalum	Ta	73	181
(technetium) ♦	Tc	43	98.9
tellurium	Te	52	128
terbium	Tb	65	159
thallium	Tl	81	204
thorium ♦	Th	90	232
thulium	Tm	69	169
tin	Sn	50	119
titanium	Ti	22	47.9
tungsten	W	74	184
uranium ♦	U	92	238
vanadium	V	23	50.9
xenon	Xe	54	131
ytterbium	Yb	70	173
yttrium	Y	39	88.9
zinc	Zn	30	65.4
zirconium	Zr	40	91.2

* Atomic masses are rounded to three significant figures

♦ All isotopes are radioactive

SAQ 5.1 (Objective 5.1)
Use the data in Tables 5.1 and 5.5 to show that the abundances of oxygen and hydrogen in the hydrosphere are consistent with their being primarily present in the form of water (H_2O), whereas this is not the case for their abundances in the lithosphere.

SAQ 5.2 (Objective 5.1)
A steelmaker wishes to evaluate several different ores for their iron content. They include: haematite (Fe_2O_3), limonite ($Fe_2O_3.2H_2O$) and magnetite (Fe_3O_4). Having chosen the richest ore, estimate the mass of iron that can be extracted from 1000 tonnes of the ore for conversion into the finished steel product.

Table 5.6 The naturally occurring elements by atomic number

Atomic number Z	Element symbol	Atomic number Z	Element symbol	Atomic number Z	Element symbol	Atomic number Z	Element symbol
1	H	24	Cr	47	Ag	70	Yb
2	He	25	Mn	48	Cd	71	Lu
3	Li	26	Fe	49	In	72	Hf
4	Be	27	Co	50	Sn	73	Ta
5	B	28	Ni	51	Sb	74	W
6	C	29	Cu	52	Te	75	Re
7	N	30	Zn	53	I	76	Os
8	O	31	Ga	54	Xe	77	Ir
9	F	32	Ge	55	Cs	78	Pt
10	Ne	33	As	56	Ba	79	Au
11	Na	34	Se	57	La	80	Hg
12	Mg	35	Br	58	Ce	81	Tl
13	Al	36	Kr	59	Pr	82	Pb
14	Si	37	Rb	60	Nd	83	Bi
15	P	38	Sr	61	Pm	84	Po
16	S	39	Y	62	Sm	85	At
17	Cl	40	Zr	63	Eu	86	Rn
18	Ar	41	Nb	64	Gd	87	Fr
19	K	42	Mo	65	Tb	88	Ra
20	Ca	43	Te	66	Dy	89	Ac
21	Sc	44	Ru	67	Ho	90	Th
22	Ti	45	Rh	68	Er	91	Pa
23	V	46	Pd	69	Tm	92	U

In the next section you will see whether the properties of the elements vary in any systematic way with their atomic numbers.

Summary

- Matter is made up of atoms, which in turn are composed of positive protons and neutral neutrons in the nucleus surrounded by a very much larger cloud of negative electrons. The number of protons determines the atomic number Z, and hence the identity of the element.
- There are 92 naturally occurring elements, of which oxygen (49%) and silicon (26%) are by far the most abundant in the accessible portions of the Earth.
- 75% of the elements are metals, 12% are gases and there are only two liquids. It is their uneven dispersion in the Earth's crust that allows them to be extracted economically.

SCIENCE FOR MATERIALS

5.3 Building the Periodic Table

This section is covered in the audiovisual activity *Building the Periodic Table* on the second side of Audiocassette 1. You should study this now in conjunction with Frames 1–8, on pp. 20–27. Note that the following terms are introduced:

alkali metals

alkaline earth metals

electron volt

group

halogens

ionization energy

noble gases

period

transition metals

The answers to the AV exercises are included with the other exercise answers at the end of this Unit.

AV1/2 Frame 1   John Newlands (left) and Dmitri Mendeleev

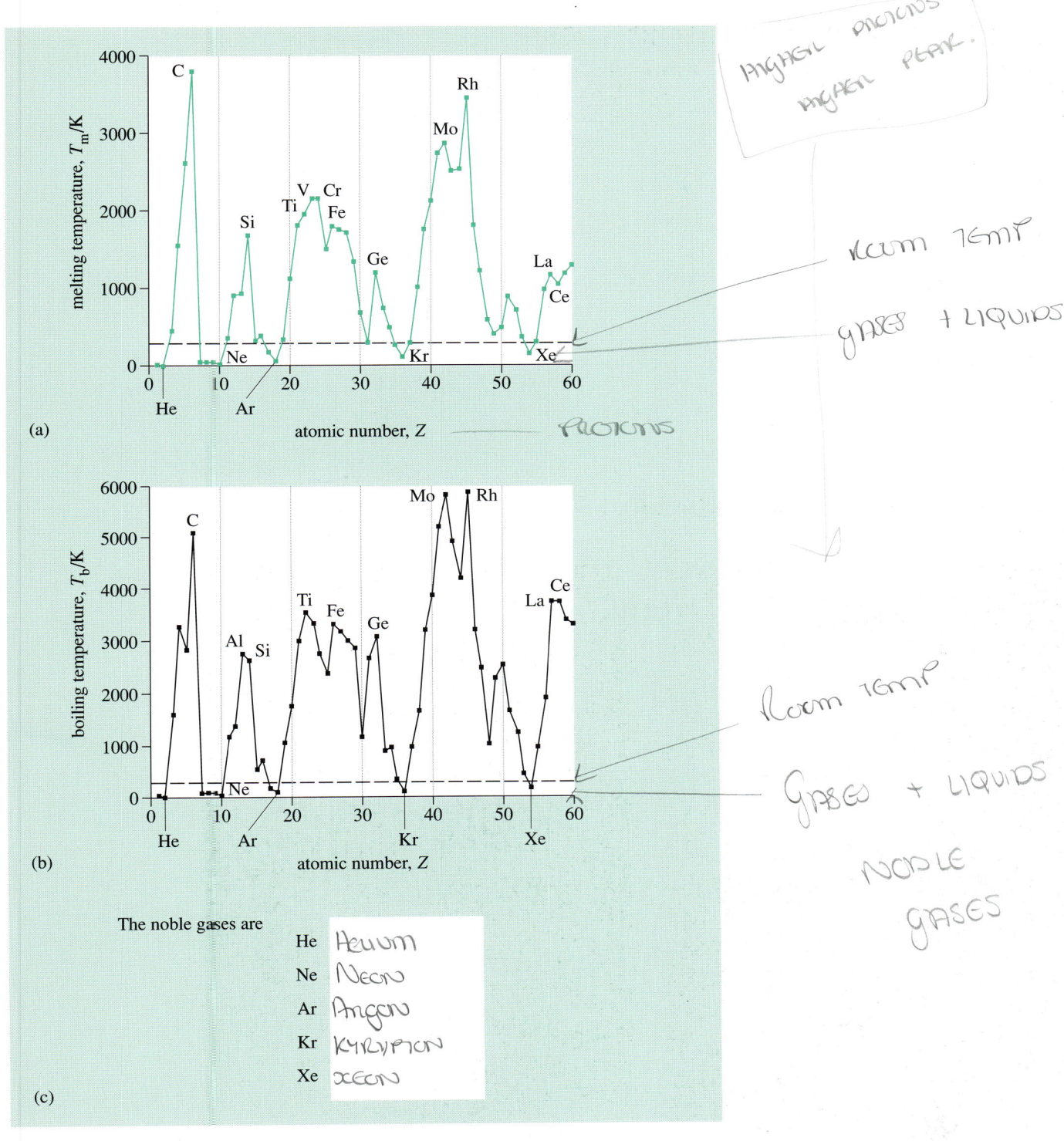

(a)

(b)

(c) The noble gases are

He Helium
Ne Neon
Ar Argon
Kr Krypton
Xe Xenon

AV1/2 Frame 2

SCIENCE FOR MATERIALS

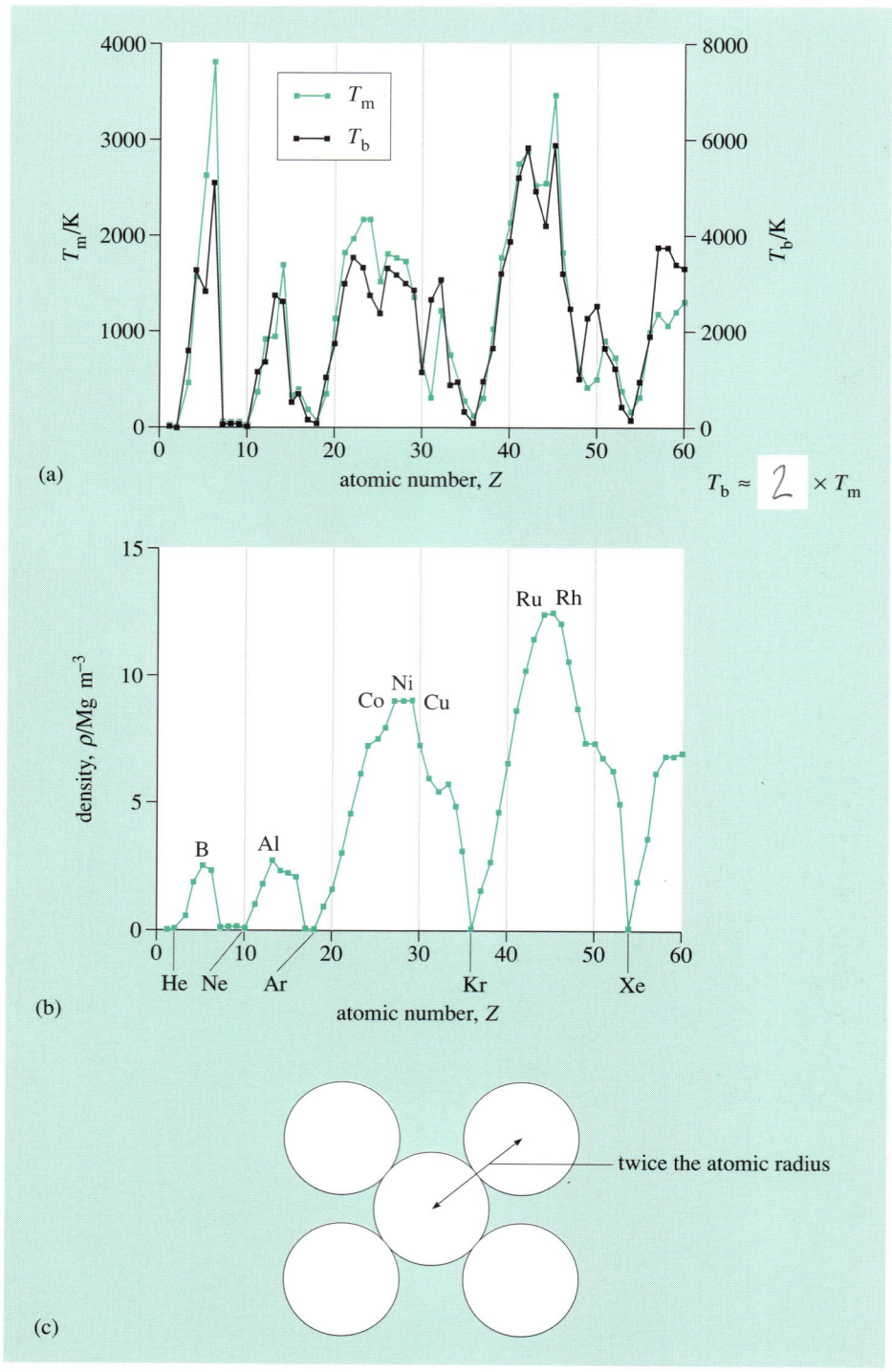

$T_b \approx 2 \times T_m$

AV1/2 Frame 3

UNIT 5 THE PERIODIC TABLE AND CHEMICAL BONDING

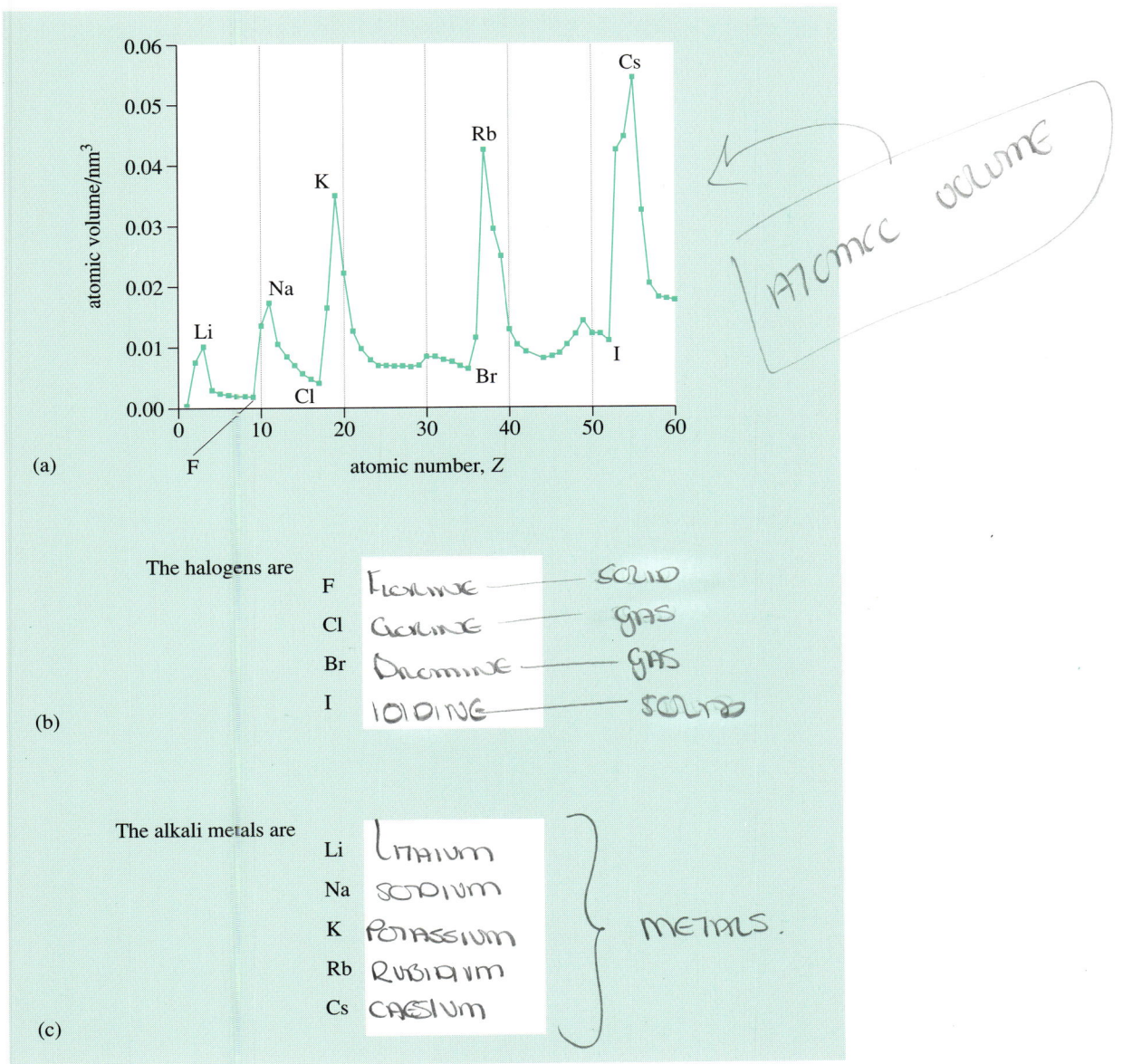

(a)

(b) The halogens are
F Florine — solid
Cl Clorine — gas
Br Bromine — gas
I Ioidine — solid

(c) The alkali metals are
Li Lithium
Na Sodium
K Potassium } metals.
Rb Rubidium
Cs Caesium

ATOMIC VOLUME

AV1/2 Frame 4

SCIENCE FOR MATERIALS

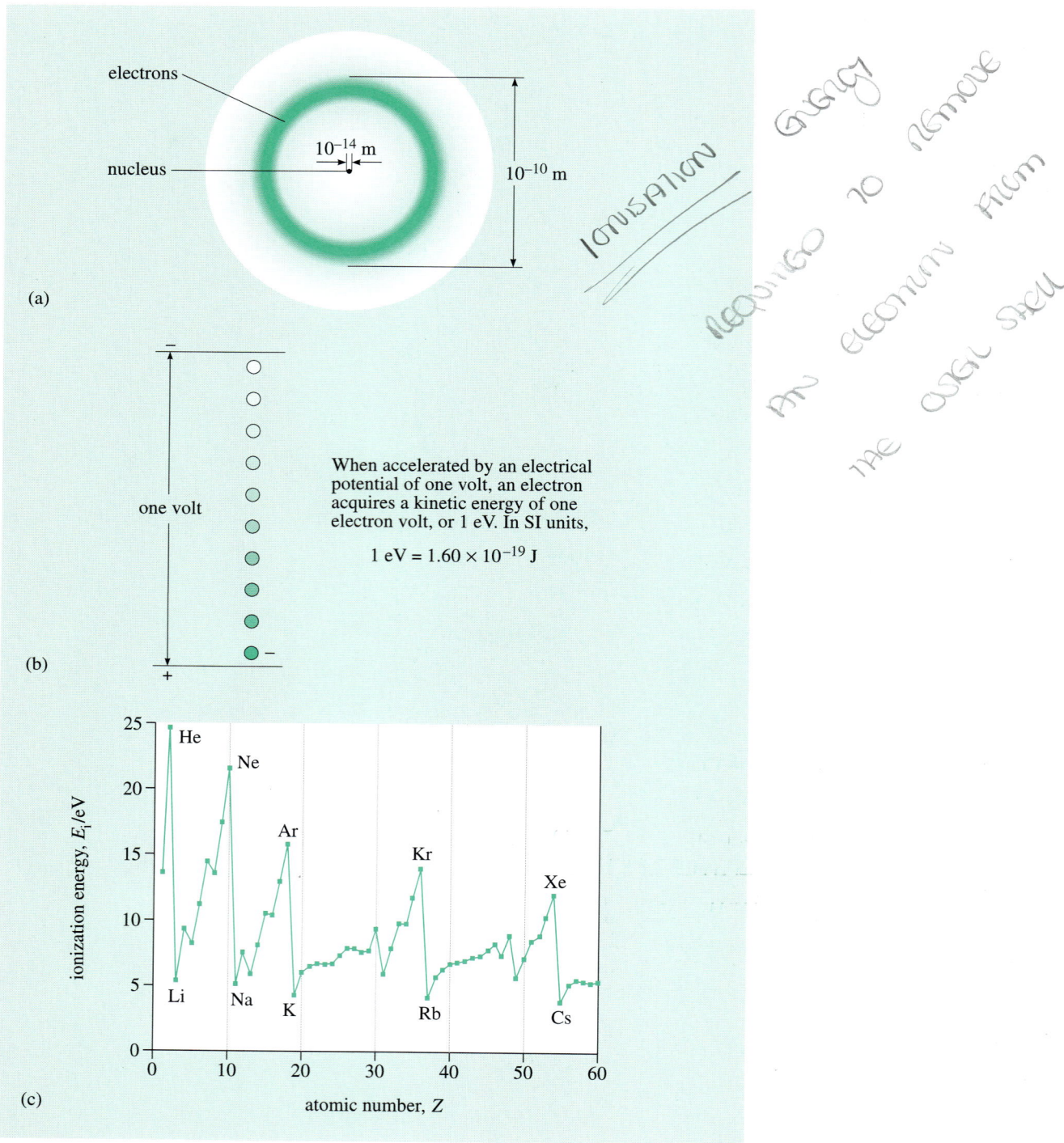

Ionisation energy required to remove an electron from the outer shell

AV1/2 Frame 5

UNIT 5 THE PERIODIC TABLE AND CHEMICAL BONDING

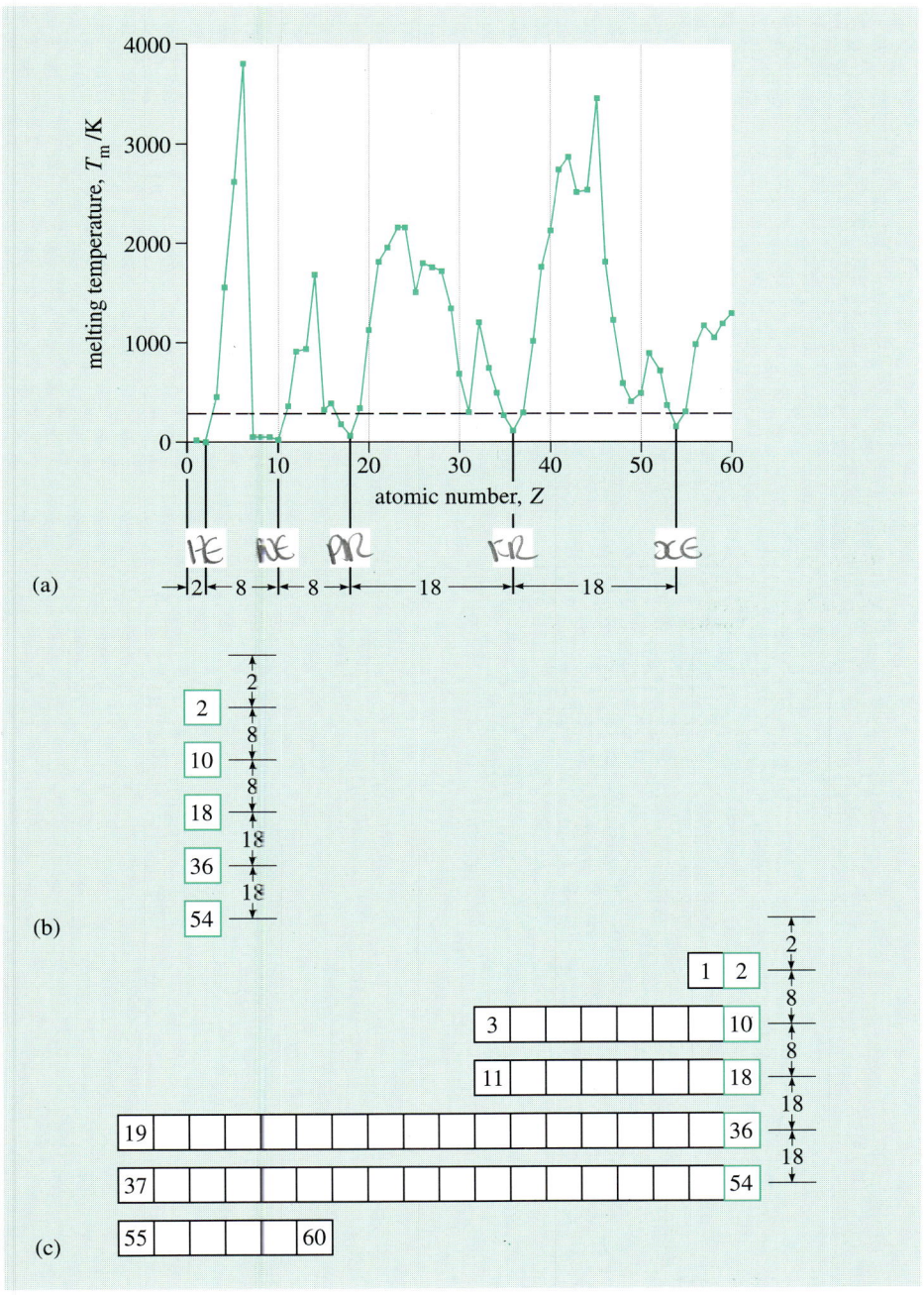

(a)

(b)

(c)

AV1/2 Frame 6

SCIENCE FOR MATERIALS

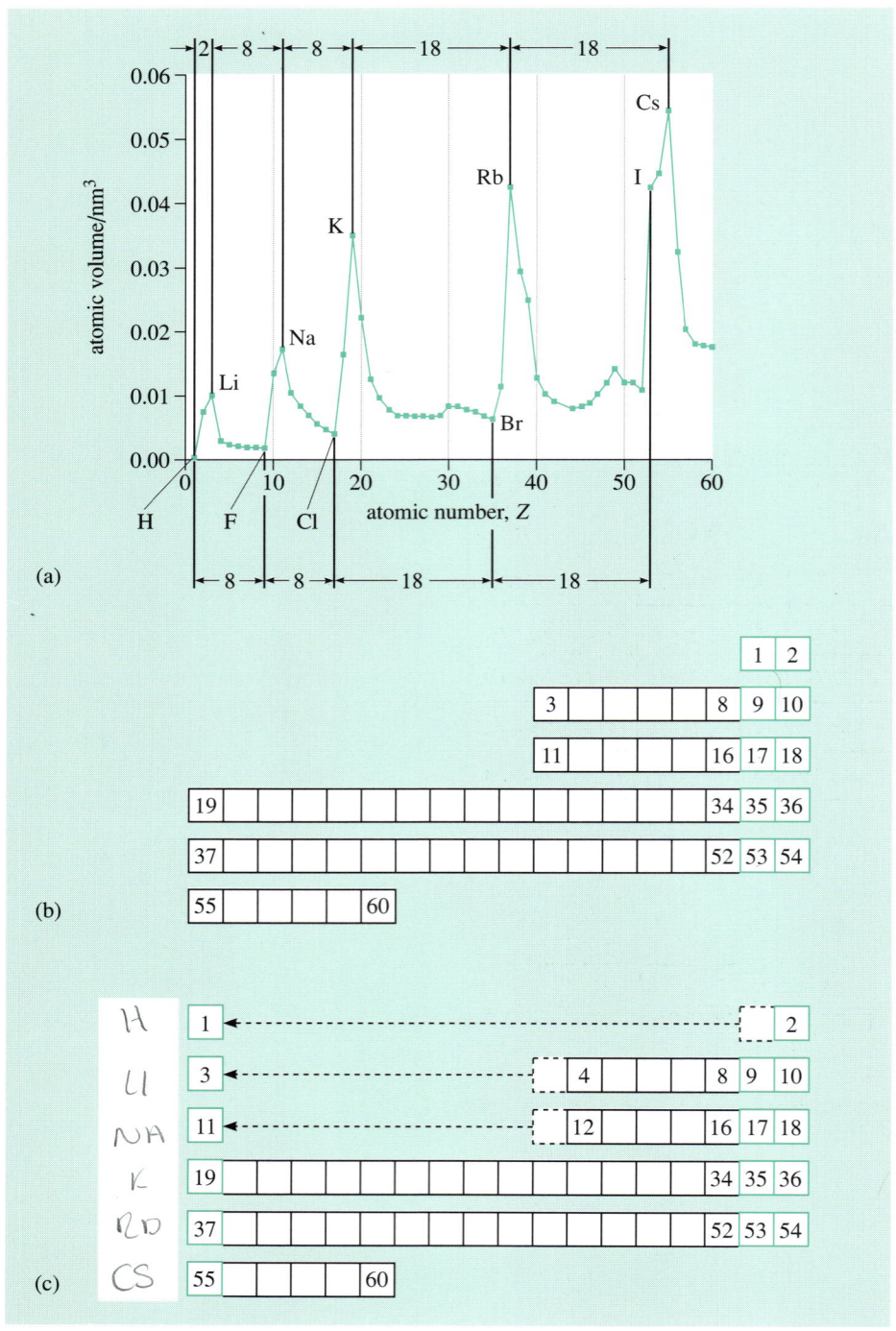

AV1/2 Frame 7

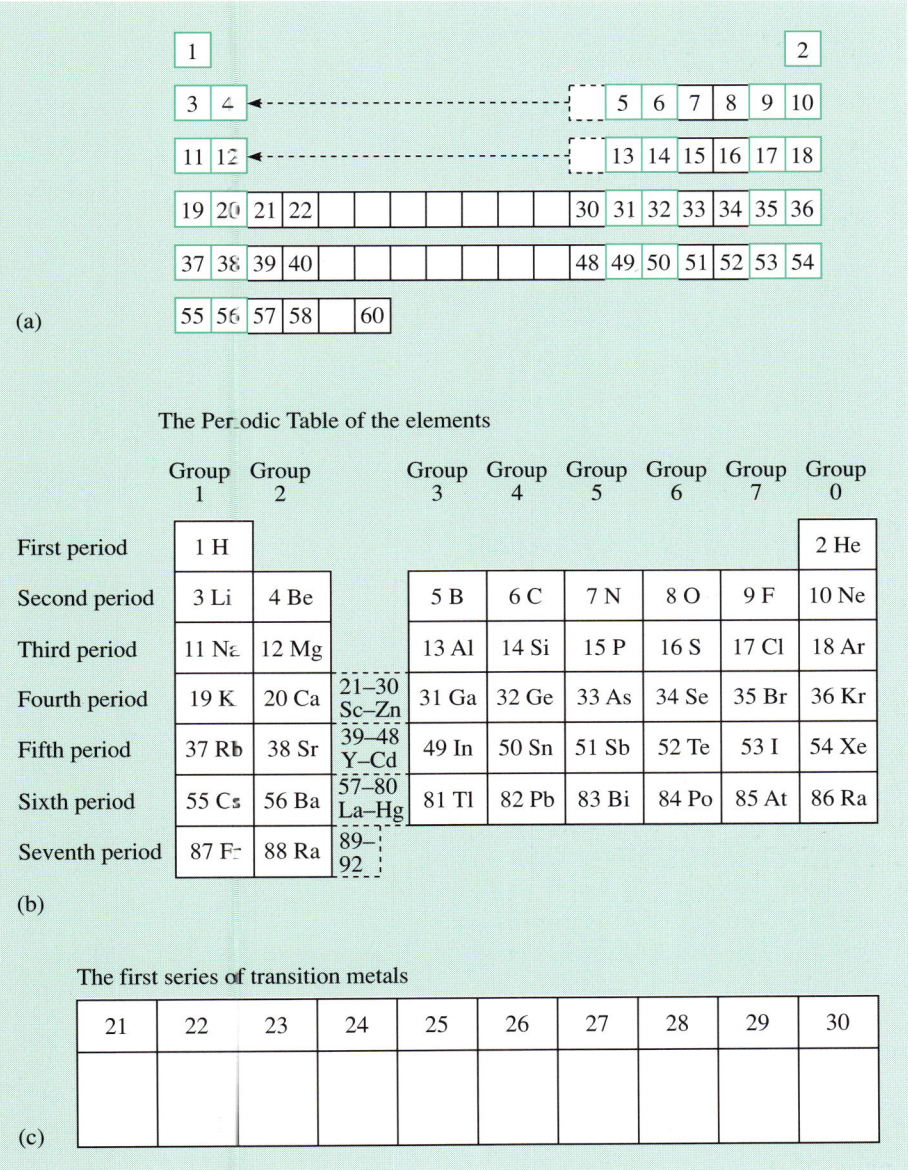

(a)

The Periodic Table of the elements

	Group 1	Group 2		Group 3	Group 4	Group 5	Group 6	Group 7	Group 0
First period	1 H								2 He
Second period	3 Li	4 Be		5 B	6 C	7 N	8 O	9 F	10 Ne
Third period	11 Na	12 Mg		13 Al	14 Si	15 P	16 S	17 Cl	18 Ar
Fourth period	19 K	20 Ca	21–30 Sc–Zn	31 Ga	32 Ge	33 As	34 Se	35 Br	36 Kr
Fifth period	37 Rb	38 Sr	39–48 Y–Cd	49 In	50 Sn	51 Sb	52 Te	53 I	54 Xe
Sixth period	55 Cs	56 Ba	57–80 La–Hg	81 Tl	82 Pb	83 Bi	84 Po	85 At	86 Ra
Seventh period	87 Fr	88 Ra	89–92						

(b)

The first series of transition metals

21	22	23	24	25	26	27	28	29	30

(c)

AV1/2 Frame 8

SAQ 5.3 (Objectives 5.2 and 5.8)
Select the appropriate words from the list below to fill the gaps in the following passage.

The Periodic Table of the _____ is based on systematic variations in physical and chemical properties with increasing _____ . Amongst these properties is the _____ , which is the energy required to remove the _____ electron from an atom. Because the energies involved at an atomic scale are very _____ , these are often measured in _____ . The noble gases have _____ ionization energies, reflecting their _____ chemical _____ . The Table is arranged into _____ vertical groups and _____ horizontal periods. Groups 1 and 2 to the left of the Table contain the _____ and _____ metals, respectively. On the right of the Table, Group 7 contains the _____ and Group _____ the _____ . Starting with the _____ period there is a gap between Groups _____ and _____ which is occupied by the _____ .

WORD LIST: alkali, alkaline earth, atomic number, eight, electron volts, elements, fourth, halogens, high, ionization energy, low, noble gases, outermost, reactivities, seven, small, three, transition metals, two, zero.

5.4 The Periodic Table and electronic structure of the elements

You've seen that the atomic number Z applies just as well to the number of electrons as to the number of protons in an electrically neutral atom. The electrons can be thought of as orbiting around the nucleus in a series of concentric spherical shells, a bit like the layers of an onion. As I mentioned in the audiovisual sequence, it's the electrons in the outermost of these shells which are the first points of contact between an atom and its environment. So it is these electrons that determine many of the atom's properties, especially those interactions with other atoms that constitute its chemical behaviour.

You also saw (or heard) that the noble gases, which form Group 0 of the Periodic Table, are distinguished by their lack of chemical reactivity compared to all the other elements. This group can thus be thought of as an 'island' of chemical stability in the Periodic Table, analogous to the minimum of potential energy in mechanical systems. In terms of electrons, the sequence of the noble gases is formed by adding successively 8, 8, 18 and 18 electrons to the initial 2 electrons of ▼Helium▲. Since it seems reasonable to assume that the most stable form of an electron shell is when it's full, it follows that the sequence 2, 8, 8, 18, 18 corresponds to the numbers of electrons required to fill the first, second, third, fourth and fifth shells respectively (as sketched in Figure 5.5).

All the other elements are less stable chemically (i.e. more reactive) than the noble gases, and have outermost shells that are less than full. To use the same analogy as above, this corresponds to their having potential

▼Helium▲

You should remember from the audiovisual exercise that helium (symbol He, atomic number 2, density at s.t.p. 0.179 kg m^{-3}) is the first of the noble gas group. Despite its relatively low abundance on Earth, it's the second most abundant element in the universe after hydrogen. It has the lowest boiling temperature (4.22 K) of any material, and, uniquely, will not freeze by cooling at atmospheric pressure. A pressure of around 30 atmospheres (≈ 3 MPa) is needed before it solidifies at just below 1 K. Liquid helium is the standard low temperature cryogenic coolant and is much used in work on superconductivity. The gas is a non-flammable alternative to hydrogen for balloons, and is used to provide an inert protective 'blanket' in arc welding and in crystal growing. It's a substitute for nitrogen in the breathing mixtures of divers and others working under high hydrostatic pressures. It's also used to pressurize liquid-fuelled rockets.

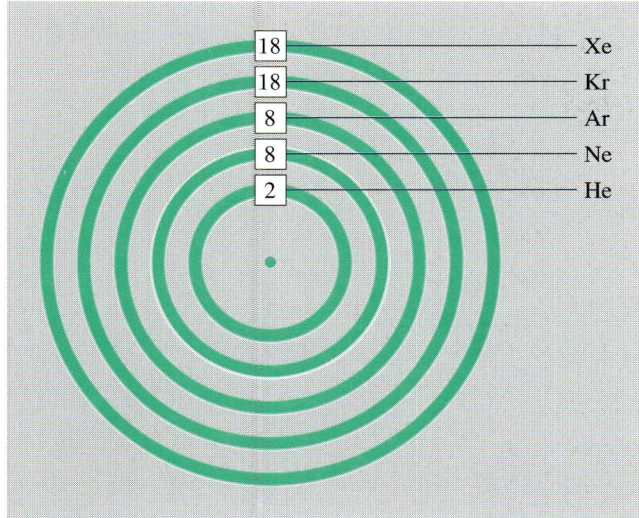

Figure 5.5 Electronic shell structures of the noble gases

energies higher than the minimum. The drive to form chemical bonds can be thought of as similar to that of mechanical systems moving to minimize their potential energy. So, for the outermost electrons in an atom, this means moving towards achieving the same shell structures as in the noble gases.

But why are such shell structures formed at all, and why should some be more stable than others?

The answers lie in the way in which classical Newtonian mechanics ceases to provide an adequate model as you go down in scale to the level of subatomic particles such as electrons. As I mentioned at the beginning of Unit 2, this is the realm of quantum mechanics. From a classical viewpoint, an electron orbiting a nucleus should emit energy in the form of electromagnetic radiation. This is because it's an atomic scale example of an electric current flowing round a conducting loop — in effect, a loop antenna (such an antenna may be more familiar as a receiving, rather than transmitting, aerial on portable TV sets). As such emission represents a continuous loss of energy, the electron should spiral in on and collapse into the nucleus. This doesn't happen — it remains in orbit in its shell.

Why it does so is accounted for by the rules of quantum mechanics. At this scale, the energy of an electron can no longer have just any value. That is, the range of allowable values is not a continuum, but rather a series of discrete steps separated by gaps, as shown schematically in Figure 5.6(a). An electron can only change energy by an amount equal to the energy spacing of one or more of these gaps. Its energy is said to be 'quantized', with each gap representing a **quantum** (plural **quanta**) of energy. An atom can absorb energy from its surroundings when one of its electrons moves from one energy state, known as an **orbital** for an electron associated with an atom or molecule, into another of higher energy (Figure 5.6(b)). It can do this only if there's room for it in the higher state, which depends on another rule of

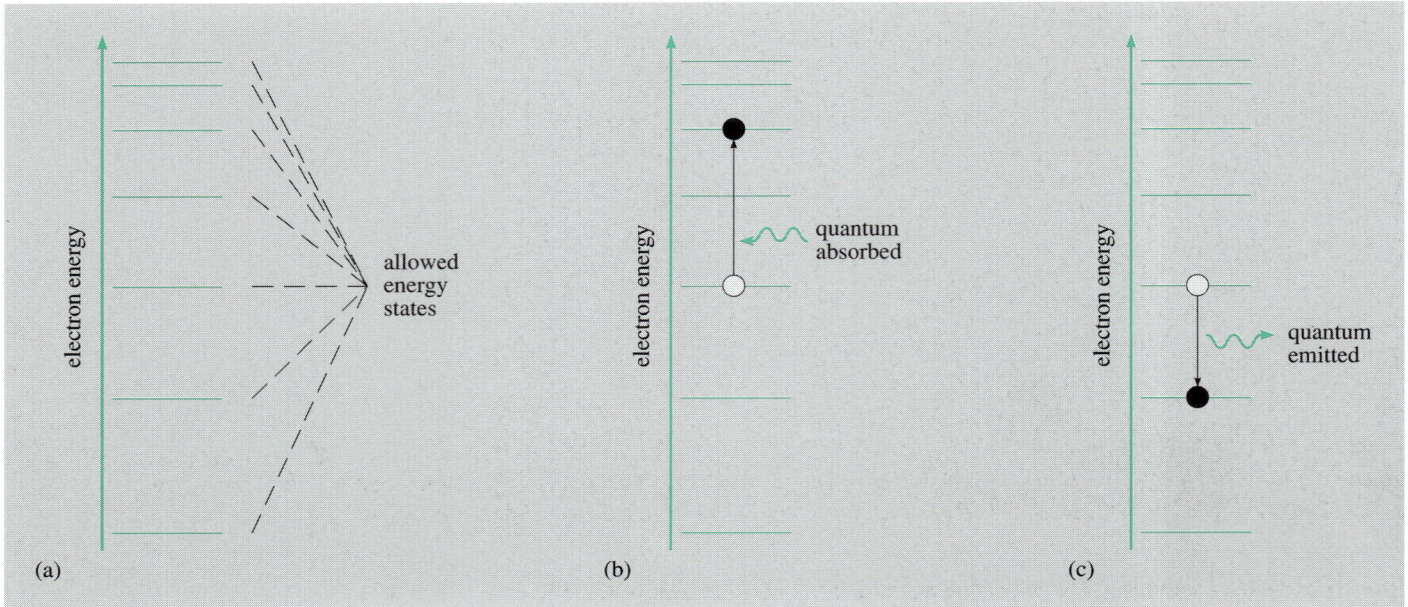

Figure 5.6 Schematic of (a) allowed energy states for an electron and (b), (c) transitions between them

quantum mechanics that you'll meet shortly. It follows, therefore, that an atom can only absorb discrete quanta of energy defined by the gaps between its energy states. An atom in such an 'excited', or higher energy, state can emit energy when the electron moves back to a lower energy state. But, once again, only as a definite quantum of energy equal to the difference in energy between the two states (Figure 5.6(c)).

So this accounts for electrons in an atom remaining in orbitals of defined levels of energy, but it doesn't tell you how many electrons there can be in each of these states, or how many states constitute a shell. To do this you need to consider more closely a set of characterizing numbers, known as **quantum numbers**.

5.4.1 Quantum numbers

The energy state of each electron in an atom is uniquely described by a set of four quantum numbers:

- the **principal quantum number**, n
- the **azimuthal quantum number**, l
- the **magnetic quantum number**, m
- the **spin quantum number**, s

Principal quantum number

The principal quantum number, n, denotes which shell an electron is in. It is always a non-zero, integral number (1, 2, 3, 4 …). Each shell has a name. The innermost shell, for which $n = 1$, is called the K-shell; the next one out is called the L-shell; and so on (Figure 5.7). Each shell contains a number of

Figure 5.7 The electronic shells and their values of principal quantum number, n

allowed states, or orbitals, so it actually represents a band of allowed values of energy. The lower the value of n, the lower is the energy of the electrons which occupy the corresponding shell.

Azimuthal quantum number

The azimuthal quantum number, l, characterizes the angular momentum of the electron, which is also quantized. But l can take values between 0 and $(n-1)$, including 0. So the K-shell, with $n = 1$, can only have $l = 0$; the L-shell, with $n = 2$, can have $l = 0$ and $l = 1$, and so on.

> **Exercise 5.6** What are the azimuthal quantum numbers that an electron can have in the N-shell?

Effectively, the azimuthal quantum number splits up the energies *within* an energy band or shell into a series of sub-shells, which are classified by a second set of letter codes.

$l = 0$ is the s sub-shell

$l = 1$ is the p sub-shell

$l = 2$ is the d sub-shell

$l = 3$ is the f sub-shell etc.

As with the principal quantum number, the higher the value of l, the higher the energy of the corresponding sub-shell.

Magnetic quantum number

The magnetic quantum number m does not normally affect the energy of an electron. It's based on the way that the energy levels of the orbitals within each sub-shell split when the atom is placed in a magnetic field. It tells you how many orbitals there are within a given sub-shell, which depends on what l is for that sub-shell. The values of m are given by the set of integers between $+l$ and $-l$, including zero. Thus, for each specified value of l, there are $(2l + 1)$ values of m, and, therefore $(2l + 1)$ orbitals.

For example, with $l = 2$ (a d sub-shell), $(2l + 1) = 5$, so there are 5 values of m and five orbitals:

$m = +2$

$m = +1$

$m = 0$

$m = -1$

$m = -2$

> **Exercise 5.7** Determine the number of orbitals in the f sub-shell of the N-shell, and list the corresponding magnetic quantum numbers.

Spin quantum number

Finally, if an electron can be regarded as a tiny solid sphere (you'll see later that it can also be modelled as a wave), then it can be thought of as spinning about its axis either clockwise or anti-clockwise. (Think of it as similar to the Earth spinning about its axis as it orbits the Sun — but it could also spin from West to East relative to the Sun.) Spin is specified by the last of the four quantum numbers, s, which can take values of $+\frac{1}{2}$ or $-\frac{1}{2}$. An important law of quantum mechanics, the **Pauli exclusion principle**, states that a maximum of two electrons is permitted in any one orbital and that when there are two, they must have opposite spins. Another way of putting this is that no two electrons in an atom can have the same set of four quantum numbers. If their values of n, l and m are all the same, they must have different values of s.

5.4.2 Allocating the electrons

You are now in a position to determine how many electrons can occupy any given set of orbitals (i.e. energy levels) associated with the electron shells around any atomic nucleus. Once the main shell (n) has been specified, then all its orbitals are uniquely set, and the number of electrons that can occupy them is determined by the Pauli exclusion principle. How this works out for the first four shells is shown in Table 5.7. The circles indicate separate orbitals, with the arrows in them representing the two electrons of opposing spin.

> **SAQ 5.4** (Objective 5.3)
> How many electrons can occupy the O-shell of an atom (where the principal quantum number is five)?

You can see that the total numbers of electrons in each shell are easy to determine given the principal quantum number, n, its relationship to l and the latter's to m. The right-hand column of Table 5.7 shows these totals to be:

K-shell ($n = 1$), only has a single sub-shell, containing one s orbital, total 2 electrons

L-shell ($n = 2$), has two sub-shells, with one s and three p orbitals, total 8 electrons

M-shell ($n = 3$), has three sub-shells, with one s, three p and five d orbitals, total 18 electrons

N-shell ($n = 4$), has four sub-shells, with one s, three p, five d and seven f orbitals, total 32 electrons.

Table 5.7 Electronic composition of shells and sub-shells

	sub-shell	s	p	d	f	
	$(2l + 1)$	1	3	5	7	total electrons per shell
n	shell	$l \leq (n-1)$	$m = -1, 0, 1$	$m = -2, -1, 0, 1, 2$	$m = -3, -2, -1, 0, 1, 2, 3$	
1	K	0	(↕)				2
2	L	0 1	(↕) (↕)(↕)(↕)			8
3	M	0 1 2	(↕) (↕)(↕)(↕) (↕)(↕)(↕)(↕)(↕)			18
4	N	0 1 2 3	(↕) (↕)(↕)(↕) (↕)(↕)(↕)(↕)(↕) (↕)(↕)(↕)(↕)(↕)(↕)(↕)			32

Do you notice anything about these totals? The first three, namely 2, 8 and 18, are just those numbers that were associated with the periodicity in properties in the previous section. Does this provide a clue to the origins of that periodicity and thus to the structure of the Periodic Table?

The answer is 'yes', but, if you remember that the earlier sequence was 2, 8, 8, 18, 18, the correspondence is not a straightforward one. To delve further you'll need to be able to handle ▼Orbital notation▲. What happens to the electrons as you go up the sequence of atoms of increasing atomic number? At each increment of Z an additional electron is added, and this joins the lowest energy orbital that is available to it, consistent with the rules you saw above.

▼Orbital notation▲

In order to talk about the distribution of electrons between different orbitals, you need to have a system of labelling and identifying particular orbitals. This is done by writing the principal quantum number n, followed by the sub-shell letter superscripted with the number of electrons in that sub-shell. For example:

$3p^6$ (spoken as 'three p six') means that in the M-shell ($n = 3$), the p sub-shell has 6 electrons in it

$1s^1$ means that in the K-shell ($n = 1$), the s sub-shell has 1 electron in it

$4d^7$ means that in the N-shell ($n = 4$), the d sub-shell has 7 electrons in it

and so on.

If you recall from Table 5.7 and SAQ 5.4 that the maximum numbers of electrons in the s, p, d and f sub-shells are 2, 6, 10 and 14, respectively, then you can see that in the first example above, the p sub-shell is full, but that in the other two they are only partially full.

It's perhaps unfortunate that this notation is similar to the index notation in maths, but $3p^2$ does *not* mean 'three times p squared'. The difference is that the sub-shell letters are set in normal (upright) type, whereas mathematical variables are italicized (e.g. $3p^2$).

You can use this notation to describe the electronic configuration of atoms. Hydrogen is $1s^1$, helium $1s^2$, lithium $1s^2 2s^1$, beryllium $1s^2 2s^2$, boron $1s^2 2s^2 2p^1$, and so on. Each sub-shell is listed in order of increasing n and l. To save having to write a long list of sub-shells, especially when you get to higher atomic numbers, the convention is to omit the inner, full shells. Thus lithium, beryllium and boron become —$2s^1$, —$2s^2$ and —$2s^2 2p^1$ respectively, the dash signifying that completed inner shells (in this case, the K-shell, $1s^2$) have been left off.

EXERCISE 5.8 A particular atom has the electronic structure —$3s^2 3p^5$. Describe its structure in words and state how many electrons it has.

The problem lies in the assumption that all sub-shells in a particular shell have a lower energy than any of those of the next shell up in the sequence. It is certainly the case that 1s has lower energy than both 2s and 2p, and these, in turn, are lower than all those in the M-shell. However, thereafter, this breaks down, as Figure 5.8 shows.

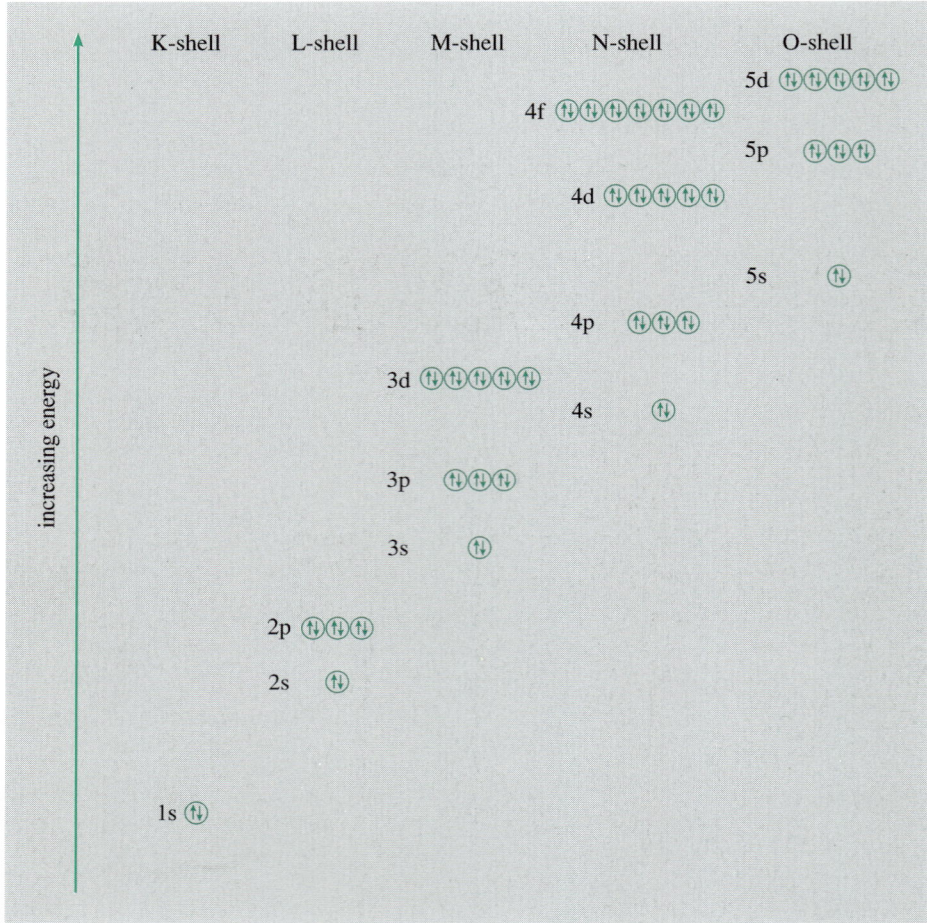

Figure 5.8 Relative energy levels of different sub-shells

Here you can see, for example, that 3d is at a higher energy than 4s, 4d is higher than 5s and 4f is higher than both 5p and 5s. How do we know this? One important tool for determining energy levels is outlined in ▼Spectroscopy▲.

So, after 3p, the sub-shells don't fill in simple order of increasing n. Rather, their overall filling sequence is as shown in Figure 5.11. Armed with this information, you can determine the electronic shell configurations of all the non-transition elements, once you know their atomic numbers.

▼Spectroscopy▲

Many previously unknown elements were first discovered using the method of spectroscopy, the study of the electromagnetic radiation emitted or absorbed by a material. One of the key facets of quantum theory is that radiation can be regarded as discrete packets (quanta) of energy. The energy E associated with any particular quantum is directly proportional to the frequency of the radiation and is given by

$$E = hf$$

where f is the frequency and h is a constant known as **Planck's constant** ($h = 6.626 \times 10^{-34}$ J s). When electromagnetic radiation interacts with matter, absorption of quanta can occur by electrons being excited to higher energy levels, as I showed earlier in Figure 5.6, provided that the energies of some of the quanta just match the gaps between the electronic energy levels. Conversely, an electron that has been stimulated to a higher energy level, whether by electromagnetic radiation or perhaps by heating the material in gaseous form or by passing an electrical discharge through the gas, can emit quanta of characteristic frequencies as it relaxes to its lowest accessible energy level (known as the 'ground state').

The result of such transitions between energy levels is the absorption or emission of a set of quanta of defined energies (or frequencies). These can be recorded and displayed as a set of lines in what is called a **spectrum** (hence the term 'spectroscopy') which shows the lines as a function of frequency. You'll find out how this is done in Unit 10. In an absorption spectrum this is a set of dark lines against a bright background, because most of the radiation passes through the material unaffected. Only those particular frequencies corresponding to the energy level differences in the constituent atoms are absorbed. In an emission spectrum it's the other way round. The energies of many of these transitions correspond to the frequencies of visible radiation. Other, higher energy ones occur in the higher frequency (shorter wavelength) region called the ultraviolet (or UV). Conversely, lower energy ones occur in the lower frequency (longer wavelength) region known as the infrared (or IR). As you'll see in Unit 10, each type of atom or molecule has its own characteristic spectrum, and so spectroscopy is a frequently used technique for identifying the constituents of materials.

The observation of absorption lines in the Sun's spectrum led directly to the discovery of 'Helium' (Greek *helios*, 'the Sun') before it was detected on Earth. 'Hydrogen' is another common element in astronomical spectra (Figure 5.9), and from its spectral lines, you can map out all the possible energy levels that its single electron can possess, depending on the degree of excitation from the ground state. Figure 5.10 shows how, starting with transitions whose quantum energies are measured from a spectrum such as those in Figure 5.9, you can build up a picture of the energy levels from which they originated. This, therefore, gives you a technique for determining the energy levels for each particular type of atom.

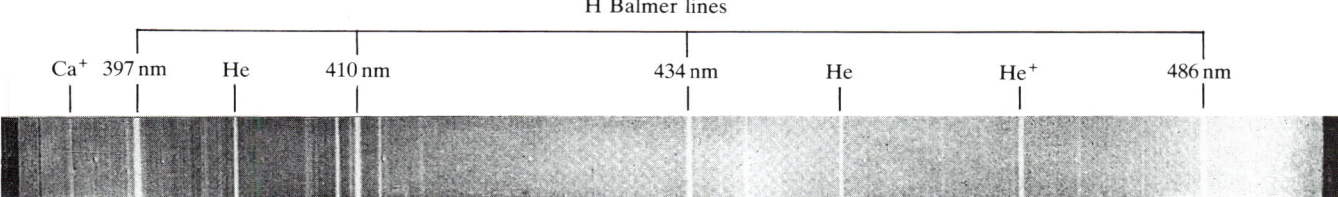

Figure 5.9 Emission spectrum of hydrogen

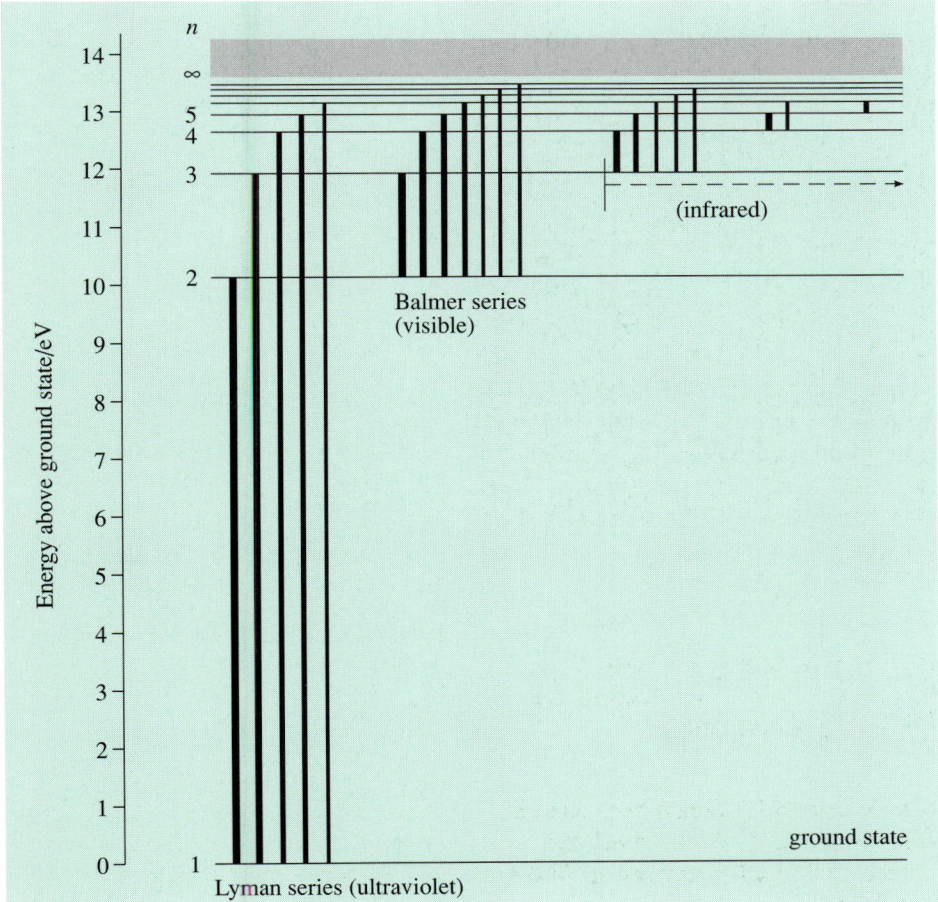

Figure 5.10 Transitions between energy levels of the hydrogen atom giving rise to different series of spectral lines (some are shown named after their discoverers) — the thicker the vertical line showing a transition is drawn, the stronger is the line

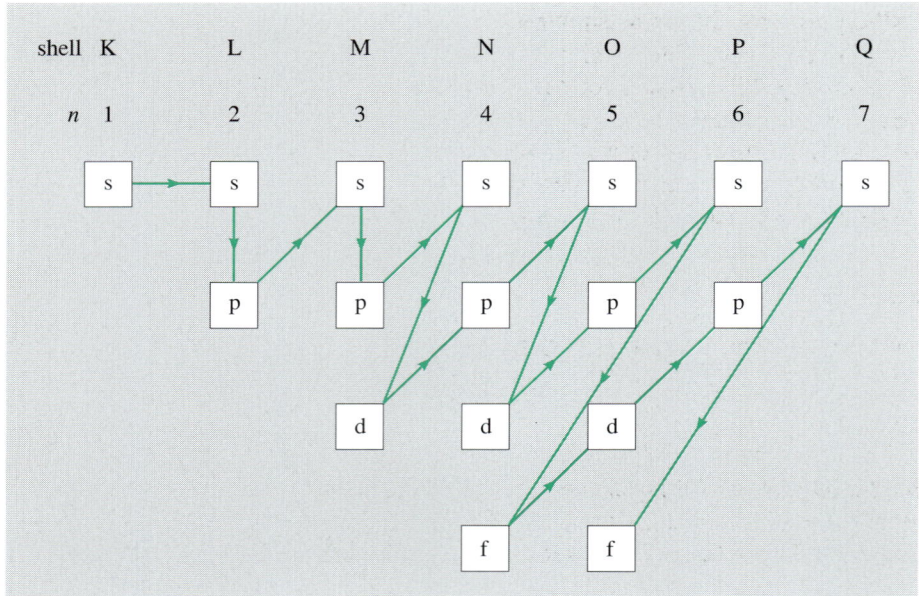

Figure 5.11 Overall sequence of filling the sub-shells with electrons

> **SAQ 5.5** (Objective 5.3)
> (a) In Figure 5.8, starting at the bottom, sketch in horizontal lines corresponding to the gaps of 2, 8, 8, 18 and 18 electrons. Use these to determine the outer shell structure of krypton.
> (b) Write down the outer shell structures of lithium ($Z = 3$), sodium ($Z = 11$), potassium ($Z = 19$) and rubidium ($Z = 37$). What do you notice about these structures, and how does this relate to the positions of these elements in the Periodic Table?

The answer tells you that the alkali metals are characterized by having an outermost shell containing a single electron in an s orbital, or —ns^1. If you did the same exercise for Group 2 of the Periodic Table, the alkaline earth metals, you'd find that they all have —ns^2 outer shells. The next group, Group 3, should then have —$ns^2 np^1$. Thereafter the p sub-shell acquires additional electrons up to the halogens in Group 7 with —$ns^2 np^5$ and, finally, you get back to the noble gases in Group 0 with —$ns^2 np^6$. This is fine for the first three periods, but ignores the presence of the transition metals between Groups 2 and 3 in the fourth and subsequent periods. If you need reminding of this, refer back to Figures (b) and (c) in Frame 8 of the audiovisual sequence in the previous section. How can you account for the presence of the transition metals? The clues to this are in Table 5.6 and Figures 5.5 and 5.8.

Let's start with the first series of transition metals, in the fourth period. Remember that the d sub-shells can hold up to 10 electrons (Table 5.6), and that the 3d sub-shell is at an energy level between those of 4s and of 4p

(Figure 5.8). This means that the order of filling goes 4s, 3d, 4p as shown in Figure 5.11. So you get —$4s^1$ (▼Potassium▲) in Group 1, followed by —$4s^2$ (▼Calcium▲) in Group 2. Then the 3d sub-shell fills, giving you the first series of transition metals, starting with —$4s^2\ 3d^1$ (namely ▼Scandium▲). Table 5.8 shows them and their electronic structures. Only when you've reached —$4s^2\ 3d^{10}$ (▼Zinc▲) can the higher energy orbitals of the 4p sub-shell start to fill, starting with —$4s^2\ 3d^{10}\ 4p^1$ (or ▼Gallium▲), which is, of course, in Group 3.

Table 5.8 Outer electron structures of the first series of transition metals

21 **Sc**	22 **Ti**	23 **V**	24 **Cr**	25 **Mn**	26 **Fe**	27 **Co**	28 **Ni**	29 **Cu**	30 **Zn**
$4s^2$	$4s^2$	$4s^2$	$4s^1$	$4s^2$	$4s^2$	$4s^2$	$4s^2$	$4s^1$	$4s^2$
$3d^1$	$3d^2$	$3d^3$	$3d^5$	$3d^5$	$3d^6$	$3d^7$	$3d^8$	$3d^{10}$	$3d^{10}$

The other thing that you can see from Table 5.8 is that the filling of the 3d sub-shell is not regular (that's why I said earlier 'all the non-transition elements'). For instance, ▼Vanadium▲ is —$3d^3\ 4s^2$, but the next in the

▼Potassium▲

You should remember potassium (symbol K, atomic number 19, relative atomic mass 39.1, density 862 kg m^{-3}) from the audio-visual exercise. It was first discovered in 1807 by Sir Humphry Davy (the British inventor of the famous Davy safety lamp). Potassium is one of the alkali metals, coming below sodium in Group 1 of the Periodic Table, and showing the high chemical reactivity of the group. Few uses have been found for it in its pure metallic form, which is silvery in appearance and soft enough to be cut with a knife. In compounds, it's an essential element for plant growth, so large quantities are used in fertilizers in the form of potash (essentially potassium oxide, K_2O).

▼Calcium▲

Calcium (symbol Ca, atomic number 20, relative atomic mass 40.1, density 1550 kg m^{-3}) is a silvery, fairly hard and reactive, alkaline earth metal, coming in Group 2 next to potassium in the Periodic Table. It's used in the chemical extraction of other metals and as a purifying agent in the processing of various alloys. Bones, teeth, shells and leaves all require calcium as an essential ingredient. Its oxide, quicklime (CaO), is extensively used in the production of mortar, plaster and Portland cement.

▼Scandium▲

Scandium (symbol Sc, atomic number 21, relative atomic mass 45.0, density 2990 kg m^{-3}) was discovered in 1876 in ores from Scandinavia, hence its name. It was first prepared in metallic form in 1937, but it wasn't until 1960 that about half a kilogram of 99% pure scandium was produced in one lump. Scandium provides the blue colour in the gemstone, beryl, but as a metal it is silvery-white, developing a yellowish-pink tinge on exposure to air. Since only about a tonne of scandium has ever been made, its uses are few, but include a window material for neutrons in nuclear reactors, and a component of the vapour in high intensity lamps.

▼Zinc▲

Zinc (symbol Zn, atomic number 30, relative atomic mass 65.4, density 7130 kg m^{-3}) is a lustrous, bluish-white, brittle metal, which is extensively used as a component of many alloys (e.g. brass, bronze, soft solder) as well as being used as a die-casting metal in its own right to make automotive components, electrical housings, toys etc. Another important application is to provide a corrosion-resistant surface layer on other metals (galvanizing), especially steels. It's an essential ingredient in the diet of animals.

▼Gallium▲

Gallium (symbol Ga, atomic number 31, relative atomic mass 69.7, density 5900 kg m^{-3}) is a silvery-white, hard, brittle metal. Its fractures look similar to those in glass. It is liquid over a larger range of temperature than any other metallic element, from just above room temperature (302.9 K) to 2676 K. One of its applications, therefore, is as the liquid in high temperature thermometry, enclosed in a quartz tube. With its low (for a metal) T_m, it's used as a component in low melting temperature alloys. It is also used as a 'dopant' to modify the electronic properties of semiconductors. Compounds of gallium are used in phosphors, and one, gallium arsenide (GaAs), is used as a semiconductor material, especially for its ability to produce laser light directly from electricity.

▼Vanadium▲

Named after the Scandinavian goddess, Vanadis, vanadium (symbol V, atomic number 23, relative atomic mass 50.9, density 6110 kg m^{-3}) is a bright white, corrosion-resistant, ductile and soft metal, first identified in 1830. Its main use is as an alloying ingredient of special steels (e.g. tool steels, spring steels), added to improve hardening.

series, ▼Chromium▲ is —3d⁵ 4s¹. Overall chromium has one more electron, but the 4s orbital has lost one electron while the 4d sub-shell complement has jumped by two. A similar thing happens when going from ▼Nickel▲ to ▼Copper▲.

There is another important point. Although each filled electronic shell in an atom is spherically symmetrical, as suggested by Figure 5.7, with the exception of the s orbitals, the individual orbitals are not. Figure 5.12 shows as an example the shapes of the three p orbitals directed along the three coordinate axes. The d and f orbitals are more complex.

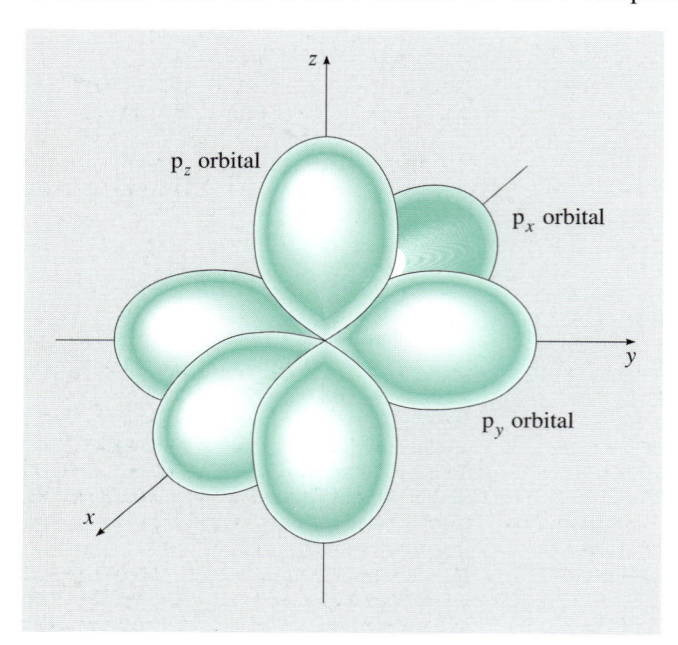

Figure 5.12 Shapes of the p-orbitals

This means that the electrons, which mutually repel each other, can take advantage of such shapes to get as far away from each other as they can within the confines of the atom, and minimize the potential energy of repulsion. Thus as more electrons are added to, say, the d orbitals, they go into separate orbitals as far as possible. It is also energetically more favourable for all those in singly occupied orbitals to have the same spins. The result is that in ▼Iron▲, for example, which has six d electrons, you get one orbital containing a pair of electrons with opposite spins, and the other four orbitals with a single electron each, all of which have their spins aligned (Figure 5.13). You'll see later that this has important consequences for magnetic behaviour.

Figure 5.13 Spins in the 3d orbitals of iron

The unevenness of filling is more frequent in the second series of transition metals, as Table 5.9 shows. Here, five of the ten elements have the 5s sub-shell only partially full (i.e. 5s¹), whilst one of them, ▼Palladium▲, actually has 5s⁰. Thus the last three elements in this series all have a full 4d sub-shell (4d^{10}), so that the additional electrons in ▼Silver▲ and ▼Cadmium▲ enter the single orbital of the 5s sub-shell.

Table 5.9 Outer electron structures of the second series of transition metals

39 **Y**	40 **Zr**	41 **Nb**	42 **Mo**	43 **Tc**	44 **Ru**	45 **Rh**	46 **Pd**	47 **Ag**	48 **Cd**
5s^2	5s^2	5s^1	5s^1	5s^2	5s^1	5s^1	5s^0	5s^1	5s^2
4d^1	4d^2	4d^4	4d^5	4d^5	4d^7	4d^8	4d^{10}	4d^{10}	4d^{10}

▼Chromium▲

Discovered in 1797, chromium (symbol Cr, atomic number 24, relative atomic mass 52.0, density 7190 kg m^{-3}) is a hard, grey, lustrous metal which polishes to a high sheen. Its main use is as an alloying ingredient in steels. Small additions improve the hardenability of the steel, and, above 12 weight %, you get stainless steel. Chromium is also electroplated onto other metals to give a hard, polished, corrosion-resistant surface. Another application is as an additive to glass which it colours emerald green. All of its compounds are coloured, hence its name, derived from the Greek *chroma*, colour.

▼Nickel▲

Nickel (symbol Ni, atomic number 28, relative atomic mass 58.7, density 8900 kg m^{-3}) was first identified in 1757 in an ore called kupfernickel, or 'Old Nick's copper', so its name means the Devil or Satan. It is a hard, silvery-white, ductile metal that is extensively used in alloying, especially in stainless steels and other corrosion-resistant alloys. It's also used in coinage, in some permanent magnets, in certain types of battery and as an electroplated surface coating of other metals.

▼Copper▲

The use of copper (symbol Cu, atomic number 29, relative atomic mass 63.5, density 8960 kg m^{-3}) dates back some 5000 years (remember the Iceman's axe in Unit 1). It's a malleable, ductile, reddish-coloured metal which is second only to silver in electrical conductivity. That's why the electrical industry is one of its major application areas. It's an important constituent of alloys such as brass and bronze and those used for coinage, and copper piping is extensively used in domestic water and central heating systems.

▼Iron▲

Used from prehistoric times, iron (symbol Fe, atomic number 26, relative atomic mass 55.8, density 7870 kg m^{-3}) is the most abundant, cheapest and most widely used metal. With up to about 2% by weight of carbon plus other alloying constituents, it forms the large variety of steels, which offer high stiffness and strength at moderate cost, and whose properties can be tailored by a combination of composition, thermal treatment and mechanical working. With between 2% and about 4.5% by weight of carbon you get the more brittle cast irons. Both forms find application in virtually every branch of engineering. Its main drawback is its susceptibility to corrosion in the presence of oxygen and water. Such corrosion can be avoided by surface coating or by alloying (especially with chromium).

▼Palladium▲

Palladium (symbol Pd, atomic number 46, relative atomic mass 106, density 12 020 kg m^{-3}) was discovered in 1803. It was not named after the theatre in London, but rather from the asteroid Pallas, which had been discovered just before (*Pallas* was the Greek goddess of wisdom). It's a metal related to platinum, is steel-white in colour, is fairly soft, and doesn't tarnish in air. It is very malleable, and can be beaten out into very thin (≈ 0.1 μm) leaf in the same way as gold, with which it is alloyed to give 'white gold'. Palladium is unusual in that it can absorb up to 900 times its own volume of hydrogen at room temperature, which is why heated palladium is used in the purification of hydrogen. It is also used in the catalytic convertors of motor cars, as well as in jewellery, watchmaking, surgical instruments and for electrical contacts and dental prostheses.

▼Silver▲

Silver (symbol Ag, atomic number 47, relative atomic mass 108, density 10 500 kg m^{-3}) has been known since around 3000 BC. It is a brilliant white, lustrous metal which, like gold, is malleable and ductile. Apart from its use in jewellery and silverware, a major application is in the emulsions of photographic film. It is also used to coat glass to make mirrors, in dentistry, in certain types of high capacity batteries, in electrical contacts and conducting paints (it has the highest electrical conductivity of any metal) and in solders and brazing alloys. Earlier, it was a major coinage metal, but its value has now outstripped the face value of the coins, so it has largely been replaced by other metals and alloys.

▼Cadmium▲

First discovered in 1817, cadmium (symbol Cd, atomic number 48, relative atomic mass 112, density 8650 kg m^{-3}) is a bluish-white, toxic metal that is soft enough to be cut with a knife. Its major use is in electroplating other metals, often to provide a basis for further coatings (e.g. enamel, chromium plating), but also finds application in solders and other low melting temperature alloys, in rechargeable batteries, and in low-friction bearing alloys.

There are two reasons for such irregular filling sequences. The first is that, as the principal quantum number increases, the corresponding energy levels get closer together. You saw this, for instance, in the energy level diagram for hydrogen in Figure 5.10. The second reason is that the energy levels don't remain at a fixed spacing as more electrons are added. Each additional proton and electron alters the electrostatic environment, resulting in minor changes to the pattern of energy levels (this helps to explain why each element has its own characteristic spectrum). In some cases the combination of these two effects leads to an overlap in the energies of different sub-shells. This is what happens with the 3d and 4s sub-shells in the first series of transition metals, and with the 4d and 5s sub-shells in the second series.

When you get to the sixth period, the first set of f orbitals, in the 4f sub-shell, comes into play. Figure 5.11 tells you that these should fill after the 6s orbital but before the 5d sub-shell. Thus, in the gap between Groups 2 and 3 in this period, you have a sequence of 14 elements corresponding to the filling of the seven 4f orbitals (each with up to two electrons), followed by a further sequence of 10 elements as the 5d sub-shell fills. The elements in the first of these sequences are called the **lanthanides** after their first member, ▼Lanthanum▲. You might also see them referred to, together with 'Scandium' and ▼Yttrium▲, as the 'rare earths'. Their outer electronic structures are shown in Table 5.10. Notice that lanthanum itself breaks the overall filling sequence by having —$6s^2\, 5d^1$ rather than —$6s^2\, 4f^1$, and that ▼Gadolinium▲ is similarly anomalous.

Table 5.10 Outer electron structures of the lanthanides

57 La	58 Ce	59 Pr	60 Nd	61 Pm	62 Sm	63 Eu	64 Gd	65 Tb	66 Dy	67 Ho	68 Er	69 Tm	70 Yb
$6s^2$	$6s^2$	$6s^2$	$6s^2$	$6s^2$	$6s^2$	$6s^2$	$6s^2$	$6s^2$	$6s^2$	$6s^2$	$6s^2$	$6s^2$	$6s^2$
$5d^1$	$5d^0$	$5d^0$	$5d^0$	$5d^0$	$5d^0$	$5d^0$	$5d^1$	$5d^0$	$5d^0$	$5d^0$	$5d^0$	$5d^0$	$5d^0$
	$4f^2$	$4f^3$	$4f^4$	$4f^5$	$4f^6$	$4f^7$	$4f^7$	$4f^9$	$4f^{10}$	$4f^{11}$	$4f^{12}$	$4f^{13}$	$4f^{14}$

Table 5.11 then shows the same information for the third series of transition metals, which includes such useful metals as gold, ▼Tungsten▲, ▼Osmium▲, ▼Platinum▲, and ▼Mercury▲ amongst its members.

Table 5.11 Outer electron structures of the third series of transition metals

71 Lu	72 Hf	73 Ta	74 W	75 Re	76 Os	77 Ir	78 Pt	79 Au	80 Hg
$6s^2$	$6s^2$	$6s^2$	$6s^2$	$6s^2$	$6s^2$	$6s^2$	$6s^1$	$6s^1$	$6s^2$
$5d^1$	$5d^2$	$5d^3$	$5d^4$	$5d^5$	$5d^6$	$5d^7$	$5d^9$	$5d^{10}$	$5d^{10}$
$4f^{14}$	$4f^{14}$	$4f^{14}$	$4f^{14}$	$4f^{14}$	$4f^{14}$	$4f^{14}$	$4f^{14}$	$4f^{14}$	$4f^{14}$

So, what you've seen in this section is how the Periodic Table, which shows groupings of the elements based on similarities in properties, is, more fundamentally, a reflection of the different electronic structures of the elements. In the next section, you'll see something of why and how elements combine with each other.

▼Lanthanum▲

Named from the Greek *lanthanein*, 'to lie hidden', lanthanum (symbol La, atomic number 57, relative atomic mass 139, density $6150\,kg\,m^{-3}$) is used in optical glasses, in carbon-arc lighting, in lighter flints, as a property-modifying additive in ceramics used in electronic applications and as a trace additive in cast iron which controls the shape of the carbon particles. It's a silvery-white, ductile and malleable metal that's soft enough to be cut with a knife.

▼Yttrium▲

Erbium, terbium, ytterbium and yttrium (symbol Y, atomic number 39, relative atomic mass 88.9, density $4470\,kg\,m^{-3}$) all get their rather odd names from Ytterby, in Sweden, where they were first discovered in ores from a quarry. Yttrium is a silvery metal that's used as an ingredient in alloys (especially of magnesium), in crystal garnets that have electronic applications as well as simulating gemstones, in the red phosphors for television screens and in some of the recent, high temperature, superconducting ceramics.

▼Gadolinium▲

First isolated in 1886, gadolinium (symbol Gd, atomic number 64, relative atomic mass 157, density $7900\,kg\,m^{-3}$) is a silvery, ductile and malleable metal, that is a very effective absorber of neutrons, and is thus used in nuclear reactors. It's also used in phosphors for television screens, as an additive to alloys which improves high temperature oxidation resistance, and as an ingredient in crystal garnets which have electronic and magnetic applications.

SAQ 5.6 (Objective 5.4)
(a) Using orbital notation, write down the general expressions for the electron configurations of the noble gases, the alkali metals, the alkaline earth metals and the halogens.
(b) Account for the presence of the transition metals and the lanthanides between Groups 2 and 3 in the Periodic Table.

Summary

- Electrons in atoms are arranged in concentric shells whose energy increases with radius, and which are identified by letter (K, L, M …) and value of principal quantum number ($n = 1, 2, 3 …$).
- Each shell contains sub-shells, designated by values of azimuthal quantum number, l, where l is the set of integers $0 \leqslant l \leqslant (n-1)$, and identified by the letters s ($l = 0$), p ($l = 1$), d ($l = 2$), etc.
- The number of available electron orbitals (or energy levels) in each sub-shell is determined by the magnetic quantum number m, an integer of value $-l \leqslant m \leqslant +l$ (including zero), so there are $(2l + 1)$ orbitals per sub-shell.
- Each orbital can accommodate up to two electrons, in which case they must have opposite spin ($s = +\frac{1}{2}, s = -\frac{1}{2}$).
- No two electrons in an atom can have the same value of all four quantum numbers (Pauli exclusion principle).
- With increasing atomic number, electrons occupy orbitals in order of increasing energy. For the first three periods up to argon ($Z = 18$), the filling pattern is regular, based on increasing n and l. However the 3d orbitals are at a higher energy than the 4s, so the filling sequence goes 4s, 3d, 4p, followed by 5s, 4d, 5p. The resulting gaps between Groups 2 and 3 as the d orbitals fill gives rise to the series of transition metals. In the sixth period, the filling becomes 6s, 4f, 5d, 6p; the 4f sequence representing the lanthanides.

5.5 Chemical bonding and periodicity

Although there are only 92 naturally occurring elements, there are literally millions of chemical compounds that can be formed between them. Given that most elements are not found free on Earth, but are combined in compounds, this argues that the chemical combination (or bonding) in these compounds leads to a state of higher stability than in the uncombined element. They are at equilibrium in a similar way to a mechanical system that has minimized its potential energy.

Exercise 5.9 What are the essential differences between a compound and a mixture? Give examples of each.

▼Tungsten▲
First isolated in 1783, tungsten or wolfram (symbol W, atomic number 74, relative atomic mass 184, density 19 300 kg m^{-3}) is a greyish-white metal that can only be worked easily when it's pure; otherwise it's very brittle. It has the highest melting temperature of all the metals, and is used for the filaments in electric light bulbs, for the heating elements of furnaces, and as an alloying ingredient in tool steels. Its carbide, WC, is very hard and is used as a cutting and grinding material.

▼Osmium▲
After iridium, osmium (symbol Os, atomic number 76, relative atomic mass 190, density 22 600 kg m^{-3}) is the second densest of the elements. It was first discovered in 1803, and is a bluish white metal that is very hard and brittle. It's difficult to fabricate and is chiefly used to make very hard alloys for pen nibs, instrument bearings, electrical contacts and the like.

▼Platinum▲
A silvery-white, corrosion-resistant, ductile and malleable metal, platinum (symbol Pt, atomic number 78, relative atomic mass 195, density 21 450 kg m^{-3}) was first discovered in about 1740. It's used in jewellery, as an electrode material, in spark plugs, in metal–glass seals, in dentistry, in thermocouples, in corrosion-resistant chemical apparatus, as the electrical elements of high-temperature furnaces, and, very importantly, as a catalyst for chemical reactions (it's finding a major use in the catalytic convertors of motor cars).

▼Mercury▲
The only metallic element that is liquid at room temperature, mercury (symbol Hg, atomic number 80, relative atomic mass 201, density 13 550 kg m^{-3}) has been known for at least 3000 years. It's used in thermometers and barometers, in discharge lamps (mercury vapour lamps), electrical switches, dental filling materials (amalgams) and some types of battery. It is toxic and needs handling with care.

SCIENCE FOR MATERIALS

So what is the nature of this bonding, and how is it achieved? You should recall from your previous studies that bonds can be classified as either stronger, **primary bonds** or weaker, **secondary bonds**. Can you identify them and their characteristics?

> SAQ 5.7 (Objectives 5.5 and 5.8)
> Select the appropriate words from the list below to fill the gaps in the following passage.
>
> The three types of primary bonding are _____ , _____ and _____ bonds. The first of these involves the transfer of _____ from one type of atom to another and the bonding, which is _____ , results from the _____ attraction between oppositely charged _____ . In the second and third, electrons are _____ between atoms. In _____ bonded materials, the shared electrons form _____ bonds between pairs of adjacent atoms. Such bonding can result in molecules with as few as _____ atoms (_____ molecules, such as hydrogen or _____) and up to the very large numbers of atoms in the giant molecules called _____ . The resulting structures can be _____ molecules, two-dimensional _____ , or _____ arrays. The third type is non-directional, and applies only to _____ , where the bonding electrons are _____ and spread out over the array of atoms.
>
> The two types of secondary bonding are _____ bonding and van der Waals' bonding. The first results from an _____ charge distribution in molecules such as _____ , whilst the second results from _____ with _____ in the charge distributions in covalently bonded materials.
>
> WORD LIST: covalent, covalently, delocalized, diatomic, directional, electrons, electrostatic, fluctuations, hydrogen, ionic, ions, linear, metallic, metals, networks, non-directional, oxygen, polymers, shared, three-dimensional, time, two, uneven, water.

So, there are basically three ways in which different atoms bond strongly one with another: ionic, covalent and metallic bonding. The last of these need only be considered briefly at this stage, but the other two warrant looking at a bit more closely.

Of all the naturally occurring elements (Table 5.2), metals are by far the most common, and are characterized by high thermal and electrical conductivities. Electrical conductivity is a measure of how easily electrons flow through the material, so in good conductors such as metals, at least some of the electrons must be able to move relatively easily under an applied voltage. These electrons are the outermost ones. Their ease of flow means that they're not bound to any single atom but rather form a 'sea' of electrons, which itself constitutes the bonding. The sea analogy helps explain why the **metallic bond**, like the ionic bond, is non-directional. The electron sea keeps the individual atoms locked in arrays which are often close-packed structures.

Before looking at ionic and covalent bonding, you might want to remind yourself of the ways of ▼Representing chemical compounds▲

▼Representing chemical compounds▲

Chemical compounds are variously represented by formulae and by models. The simplest formula is the **empirical formula**, which gives you the ratios of the atoms present in a compound. So, NaCl tells you that sodium chloride is based on equal numbers of sodium and chlorine atoms and, as you saw earlier (SAQ 5.1), water, with the empirical formula H_2O, has twice as many hydrogen atoms as oxygen atoms. (Note the convention of writing the '2' as a subscript; 2HO means something quite different.)

Although empirical formulae are fine for ionic compounds, which don't form discrete molecules, they fall down when it comes to covalently bonded molecules. For instance, the empirical formula CH_2O can represent CH_2O (formaldehyde or methanal, a gas used in aqueous solution as a disinfectant), $C_2H_4O_2$ (acetic or ethanoic acid, the basis of vinegar), $C_3H_6O_3$ (lactic acid alias 2-hydroxy-propanoic acid, which builds up in muscles after strenuous exercise) and so on. All of which have the same ratios of carbon, hydrogen and oxygen. The last three formulae are the **molecular formulae** for the compounds in question, which gives you the ratios of the atoms in each individual molecule. The molecular formula can also be written in a more informative way to tell you how the atoms are grouped. Formaldehyde becomes H.CHO and acetic acid $CH_3.CO(OH)$, where CHO is the aldehyde group, CH_3 the methyl group and CO(OH) the carboxylic acid group. The dots are used to separate each group and the brackets round the OH signify that it too is joined directly to the carbon atom. The CHO and CO(OH) groups are known as **functional groups** because they determine the class of chemical behaviour of the molecule.

A **structural formula** takes you one stage further. It shows you the order and nature of the bonds connecting the groups in a molecule. You'll see shortly that, depending on the number of electrons forming a covalent bond (2, 4 or 6), the bond is classed as single, double or triple, and these are represented in structural formulae by a single line ($-$), a pair of parallel lines ($=$) and three parallel lines ($\equiv$) respectively. Figure 5.14 shows you the structural formulae for formaldehyde and acetic acid as examples.

Remembering that covalent bonds are directional, it would clearly be helpful to have a way of showing their arrangement in space. There are two models used to do this. The first is the 'ball-and-stick' model, which, as its name suggests, represents atoms by balls and the bonds between them by lines or sticks. Although the bond angles are the true ones and their lengths are often shown to scale, the balls are smaller than scale atoms would be, so that the bonds can more easily be seen. You'll be able to experiment with ball-and-stick models at Summer School. The wire-frame model is even sparser, depicting only the bonds and their directions, with no model atoms. This is more useful for revealing the patterns of bonding in a large array, such as an extended crystal lattice. Figure 5.15 shows a ball-and-stick model and a wire-frame model for acetic acid.

Figure 5.14 Structural formulae of formaldehyde and acetic acid

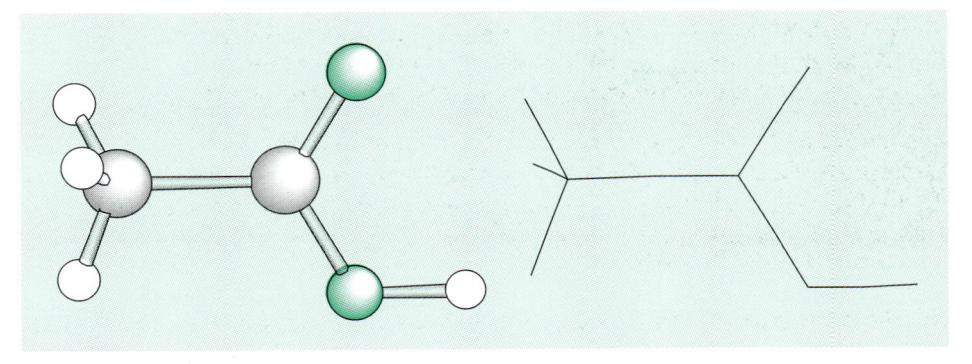

Figure 5.15 Ball-and-stick and wire-frame models of acetic acid

5.5.1 Ionic bonding

I said earlier that it's the outermost electrons that are responsible for bonding. One way the outermost electrons of two different atoms interact is simply by electrons transferring from one atom to the other. Since the starting atoms are electrically neutral, the result is two charged particles, or **ions**. The loser atom becomes a positively charged ion, known as a **cation** because of how it would move in an electric field if it were free to do so — see Figure 5.16. The gainer becomes a negatively charged one, known as an **anion** for similar reasons. As a result, there is an electrostatic attraction between these opposite charges that constitutes the **ionic bond**.

But what governs which atoms donate and which receive electrons, and what determines the number of electrons gained or lost?

In general, metals lose electrons to become positively charged cations and non-metals (e.g. oxygen, hydrogen, the halogens) gain electrons to become negatively charged anions. The number of electrons transferred is determined by a property known as the **valency** of an atom.

I also said earlier that the most stable electronic configurations are those of the noble gases. The drive for bonding is a drive towards greater stability. The valency of an element is determined by the number of electrons its atom needs to gain or lose in order to achieve a more stable electronic structure.

You can obtain this by reference to the electronic configurations that you saw in the previous section, so this, in turn, links periodicity and chemical bonding.

SAQ 5.8 (Objectives 5.3 and 5.6)
What changes in their electronic structure would be needed for elements in Group 2 and in Group 7 of the Periodic Table to achieve a noble gas configuration? What types of ion do they form and what are their expected valencies?

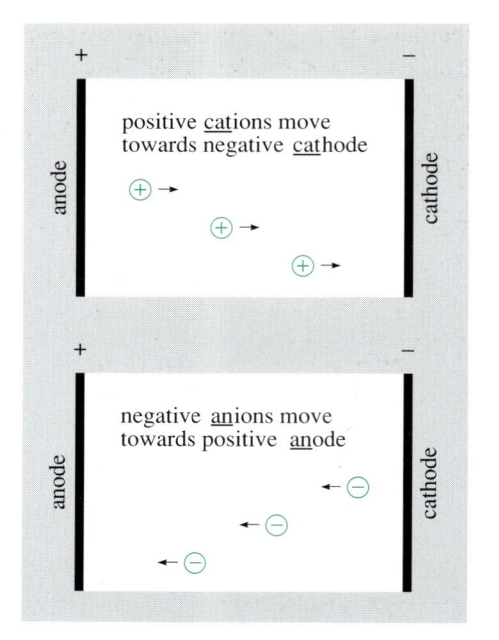

Figure 5.16 Movement of positive ions (cations) towards the negative electrode (cathode) and negative ions (anions) towards the positive electrode (anode) in an electric field

Table 5.12 shows the number of electrons transferred and valencies for a selection of anion formers. This shows that, for example, hydrogen or fluorine compounds will have single negative charges on their anions, which are written as H^- and F^-, respectively.

Table 5.12 Electron transfer and valency of anion formers

1 electron gained (monovalent)	2 electrons gained (divalent)	3 electrons gained (trivalent)	4 electrons gained (quadrivalent)
hydrogen	oxygen	nitrogen	carbon
fluorine	sulphur	phosphorus	silicon
chlorine	selenium	boron	
bromine			
iodine			

In the second column, you'd expect oxygen and sulphur to form doubly-charged O^{2-} and S^{2-}, and in the third and fourth columns, to get N^{3-}, P^{3-}, C^{4-} and Si^{4-} from nitrogen, phosphorus, carbon and silicon.

A similar list can be made for the metals, although the situation is rather more complex, with many metals showing multiple valency. Their ions can have different values of charge in different compounds. 'Iron' and 'Copper' on p.39 provide examples of this. Other metals, however, form only one type of cation, and four typical groups are shown in Table 5.13. Note that I've included hydrogen in the first column of this table as well (remember that, in *Building the Periodic Table*, hydrogen could have been grouped either with the halogens or with the alkali metals). If you compare the columns in Tables 5.12 and 5.13 with the groups of the Periodic Table, you'll see that there is a strong correspondence between them.

Table 5.13   Electron transfer and valency of cation formers

1 electron lost (monovalent)	2 electrons lost (divalent)	3 electrons lost (trivalent)	4 electrons lost (quadrivalent)
hydrogen	beryllium	aluminium	zirconium
lithium	magnesium	gallium	hafnium
sodium	calcium	indium	
potassium	strontium	scandium	
rubidium	barium		
caesium	zinc		
silver	cadmium		

So, based on Table 5.13, you'd expect to get the cations H^+, Na^+, Mg^{2+}, Al^{3+} and Zr^{4+}, for example. Given the simple rule that, like atoms, compounds are electrically neutral, you're now in a position to predict the empirical formulae of ionic compounds. Thus magnesium oxide (see ▼Naming ionic compounds▲) should consist of equal numbers of ions of Mg^{2+} and O^{2-}, since the 2+ neutralizes the 2−, so its formula is MgO. Barium hydride requires two negative hydride ions (H^-) to balance its divalent state, so you get the formula BaH_2. Hydrogen, uniquely, features in both Tables 5.12 and 5.13, in line with what you saw of its two possible positions in the Periodic Table. Most of its compounds are covalent, but when it loses an electron, you get H^+, which is just a bare proton, with an intense, positive electrostatic field. This is very reactive, and in aqueous solution it combines with a water molecule to give what's known as the **hydroxonium ion**, H_3O^+. It forms the basis of such acids as hydrochloric acid, which is nominally HCl, but is actually a solution in water of H_3O^+ and Cl^-. But, as you've seen, hydrogen can also gain an electron to become the H^- anion, and so form ionic hydrides.

▼Naming ionic compounds▲

By convention, ionic compounds are named with the positive cation (usually a metal) first, followed by the negative anion. The cation retains the name of the element. Where the anion is derived from a single element, its name, though based on that of its parent element, is terminated with '-ide'. Hence oxide, chloride, hydride, arsenide, carbide, nitride, and so on. This also applies to the names of classes of compound such as the alkali halides. (It's a bit unfortunate that groups of elements such as the lanthanides and actinides also share the same suffix.) Anions can also be formed by groups of elements, as you'll see shortly. Some of these (e.g. cyanide, CN^-, or hydroxide, OH^- retain the '-ide' suffix. Others (e.g. nitrite, NO_2^-, carbonate CO_3^{2-}) have other suffixes.

Exercise 5.10   What are the empirical formulae of
(a) silver oxide
(b) calcium chloride
(c) aluminium oxide?

The ionic bond is not just a simple matter of two single atoms (e.g. sodium and chlorine) coming together, exchanging an electron and then bonding to one another by the resulting electrostatic force. This would imply the existence of discrete, ionically bonded molecules. As I mentioned in

'Representing chemical compounds' on p. 43, such entities don't exist. Work is needed to remove an electron from an atom against the attraction of the positively charged nucleus. Work is also needed to add an electron to a neutral atom against the repulsion of its existing electrons. The reduction in potential energy in forming such a single ionic bond is insufficient to compensate for these two work terms, so it doesn't happen. Ionic bonds only form when each ion can be surrounded by an array of ions of opposite charge, each of which in turn being surrounded by an array of ions whose charge is opposite to theirs. This optimizes the balance between electrostatic attraction and repulsion and leads to an overall reduction in potential energy.

The net result is a solid with an extended crystal structure, or **superlattice**, satisfying the above condition. Since the electrostatic force acts equally in all directions, the bonding is non-directional. When an atom loses an electron to become a cation, the excess positive charge on the nucleus draws the remaining electrons in more closely, so a cation is smaller than its parent atom. The converse applies with anions: the extra negative charge increases the mutual repulsion of the electrons, so they spread out more, making the anions larger. The relative size of the two types of ion controls the crystal structure in the solid state. The smaller cations fill the interstices in an array of the larger anions (Figure 5.17), so that the necessary balance between electrostatic attraction and repulsion is achieved.

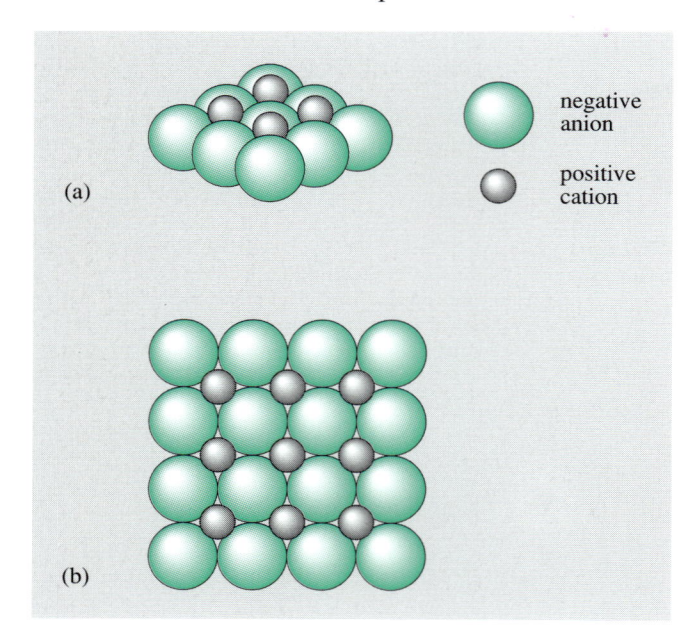

Figure 5.17 Array of positive and negative ions in a layer of an ionic crystal, (a) perspective, (b) plan view

5.5.2 Covalent bonding

The other principal way in which the outer electrons of elements form bonds, either with like atoms or with different elements, is by being *shared*. Two shared electrons constitute a single covalent bond, four electrons a double bond and six electrons a triple bond. The idea of valency still applies, but the two, four or six shared electrons orbit about both the nuclei involved (for two elements bonding together). In contrast to ionic and metallic bonds, **covalent bonds** are highly directional, since the electrons are localized in space between the pair of atoms. Covalent bonding leads to the formation of molecules.

What kinds of compounds are covalent?

The simplest examples are those where identical elements are involved. One example is oxygen. It forms a diatomic molecule, written as O_2, in which two oxygen atoms bond together by sharing electrons. Its structural formula is written as O=O where each single line represents *two* shared electrons (remember from Table 5.12 that each oxygen atom can donate *two* electrons), so two lines represent a total of four shared electrons.

> **Exercise 5.11** Given that hydrogen, fluorine and chlorine all form diatomic molecules, write the structural formulae for their normal gaseous states.

What happens in covalent bonding is that the atomic electron orbitals combine to form **molecular orbitals** at a lower energy. Figure 5.18 shows this schematically for hydrogen. The directionality of the bond comes from the molecular orbitals being concentrated along the line of centres of the atoms.

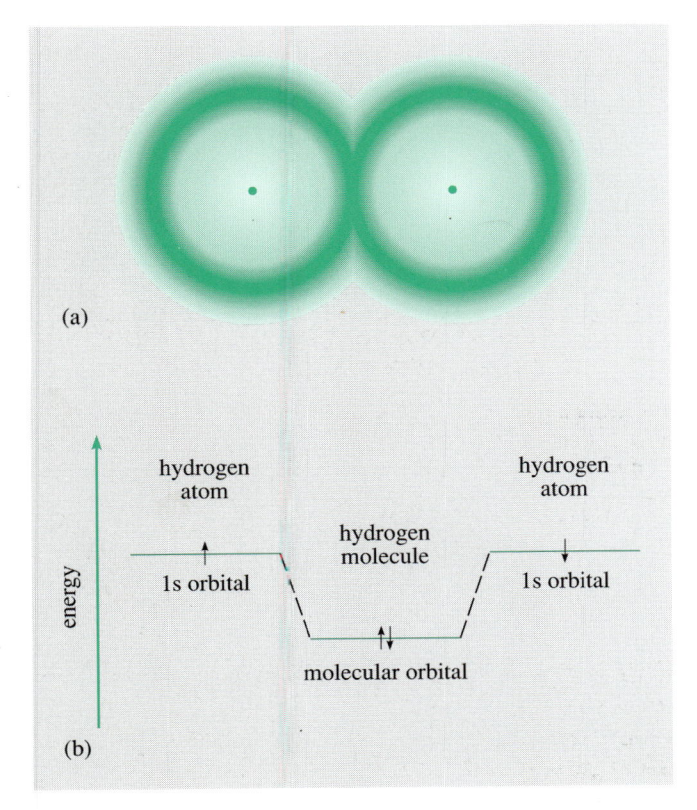

Figure 5.18 Schematics of (a) hydrogen molecule, (b) lowering of energy in molecular orbital

Pairs or groups of dissimilar elements, especially non-metals, can also share electrons. Water (H₂O) is one example. With oxygen divalent and hydrogen monovalent, its structural formula is H—O—H, showing that two single bonds between oxygen and hydrogen are formed.

One of the strongest covalent bonds is that formed between carbon atoms:

$$\begin{array}{c} \uparrow \quad \uparrow \\ \leftarrow C - C \rightarrow \\ \downarrow \quad \downarrow \end{array}$$

Each carbon atom has three unpaired electrons available to bond with other elements as represented by the three arrows on each atom. Due to electrostatic repulsion between the four molecular orbitals associated with any one carbon atom, the bonds are directed in space as far from each other as possible. These orbitals are the result of a combination of atomic s and p orbitals, and are known as **sp³ hybrid** orbitals. If you imagine the carbon atom sitting in the centre of a tetrahedron (Figure 5.19), this leads to the bonds pointing towards the corners (or vertices), and produces the characteristic shapes of such carbon compounds.

(a) tetrahedral shape   (b) shared electrons   (c) orbitals

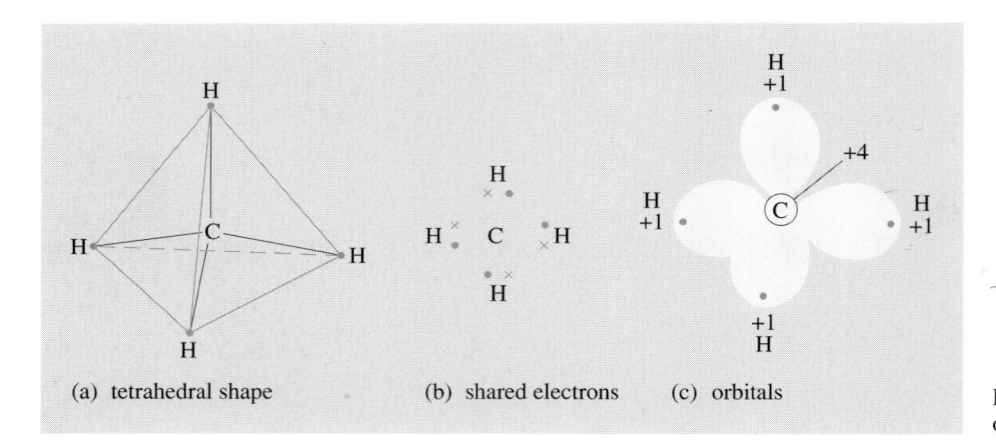

Figure 5.19   Tetrahedral arrangement of carbon bonds

If the element linked to the spare bonds on the pair of carbon atoms in the structural formula at the top of this page is hydrogen, which is monovalent, you get ethane

$$\begin{array}{cc} H & H \\ | & | \\ H-C-C-H & \quad C_2H_6 \\ | & | \\ H & H \end{array}$$

It's commonly found in natural gas. Figure 5.20 shows a ball-and-stick model of ethane.

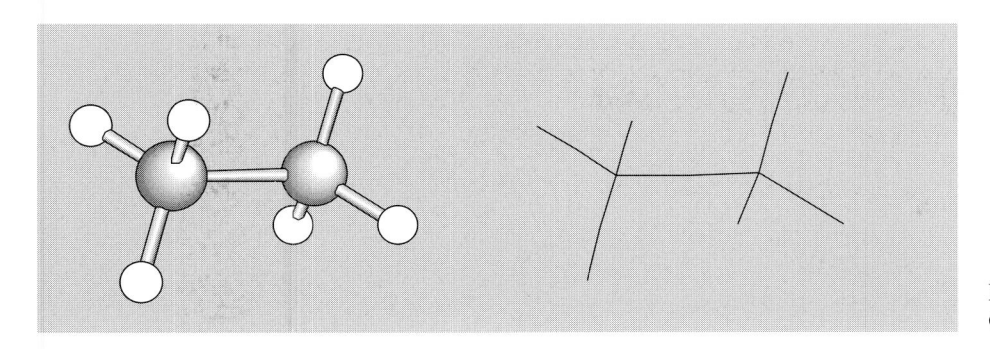

Figure 5.20 Ball-and-stick model of the ethane molecule

The important feature of ethane is that it's just one member of a much larger family (called the linear **hydrocarbons** — compounds of hydrogen and carbon) in which a whole series of similar molecules can be formed by adding successive CH_2 groups.

$$\begin{array}{c} H\ \ H\ \ H \\ |\ \ \ |\ \ \ | \\ H-C-C-C-H \\ |\ \ \ |\ \ \ | \\ H\ \ H\ \ H \end{array} \quad C_3H_8 \text{ (propane)}$$

$$\begin{array}{c} H\ \ H\ \ H\ \ H \\ |\ \ \ |\ \ \ |\ \ \ | \\ H-C-C-C-C-H \\ |\ \ \ |\ \ \ |\ \ \ | \\ H\ \ H\ \ H\ \ H \end{array} \quad C_4H_{10} \text{ (butane)}$$

$$\begin{array}{c} H\ \ H\ \ H\ \ H\ \ H \\ |\ \ \ |\ \ \ |\ \ \ |\ \ \ | \\ H-C-C-C-C-C-H \\ |\ \ \ |\ \ \ |\ \ \ |\ \ \ | \\ H\ \ H\ \ H\ \ H\ \ H \end{array} \quad C_5H_{12} \text{ (pentane)}$$

and so on. This process can be continued virtually indefinitely: the end-result is polyethylene, one of the class of materials called polymers, which can have thousands of carbon atoms joined together to make its chain molecules.

> **Exercise 5.12** Write out the structural formula of the simplest possible hydrocarbon (i.e. the one containing the smallest number of single carbon–hydrogen bonds).

A pair of carbon atoms can also link via double and triple bonds and still retain the noble gas structure between them. With hydrogen as the substituent atoms again, you get two new compounds:

$$\begin{array}{cc} H & H \\ \diagdown & \diagup \\ C=C \\ \diagup & \diagdown \\ H & H \end{array} \qquad \text{ethylene (or ethene)}, C_2H_4$$

$$H-C\equiv C-H \qquad \text{acetylene (or ethyne)}, C_2H_2$$

These bonds involve hybrid molecular orbitals which differ from the sp^3 — you'll meet these in more detail later. As you might imagine, such double- and triple-bonded molecules are more reactive than their single-bonded cousins. The former are known as **unsaturated** molecules because you can add further substituents to each carbon atom by using one of the 'spare' bonds between the carbon atoms. (Molecules which have just single bonds between the carbon atoms are called **saturated**). Acetylene is a very reactive gas which is used, for example, in oxyacetylene welding. Ethylene is used to make polyethylene (as you might have guessed from its name), losing its double bonds in the process to give the saturated hydrocarbon structure that you saw above.

By substituting other species for hydrogen, you get other molecules. For instance, substituting a methyl group (CH_3) gives you:

$$\begin{array}{cc} H & CH_3 \\ \diagdown & \diagup \\ C=C \\ \diagup & \diagdown \\ H & H \end{array} \qquad \text{propylene (or propene)}, C_3H_6 \text{ or } C_2H_3.CH_3$$

This is the starting point for polypropylene.

Another example: substitute chlorine for one of the hydrogens and you get:

$$\begin{array}{cc} H & Cl \\ \diagdown & \diagup \\ C=C \\ \diagup & \diagdown \\ H & H \end{array} \qquad \text{vinyl chloride (or chloroethene)}, C_2H_3.Cl$$

This is also known as vinyl chloride monomer (or VCM), which you might have seen in the news a few years ago because of concern over its carcinogenic effects. VCM is the starting point for poly(vinyl chloride), better known as PVC, widely used in pipe-work and guttering, for glazing, as a leather substitute in upholstery, as the insulation of electrical wires and cables etc. If you continue substituting onto the same carbon atom, you reach a stage where you have to take ▼**Symmetry**▲ into account.

UNIT 5 THE PERIODIC TABLE AND CHEMICAL BONDING

▼Symmetry▲

We should all be familiar with symmetrical objects: our right hands are almost a mirror image of our left hands (Figure 5.21). The idea of symmetry is very general and is particular useful for comparing the shapes of similar but not identical objects or shapes in space. For the sp^3 bonded carbon atom, several possibilities arise. When one hydrogen atom is substituted by a different atom (X), then translating the mirror image back onto the original structure produces perfect identity since there are three identical hydrogen atoms in the molecule. If one of the remaining hydrogen atoms is replaced by a second different atom (Y), the result is another identity — simply rotate the mirror image through 180° about the vertical C—X bond, and translate back onto the original structure, as shown in Figure 5.21(c).

However, a completely different situation arises if a further hydrogen atom is replaced by a third substituent (Z). This situation is shown in Figure 5.21(d). Whatever operations are performed on the mirror image of the molecule, it will *never* be identical to the original. In other words, the mirror image is another, distinct molecule, and it has distinctly different properties from the original. This effect only occurs when all four substituent groups are different, when the central atom is called **asymmetric**. The two different configurations are known as the left- (L-) and right- (R-) handed **stereoisomers** of the molecule respectively. The effect is not limited to carbon, but can occur on any atom with four or more bonds (e.g. silicon).

Stereoisomerism occurs widely in natural molecules, where all asymmetric carbon atoms are found to be left-handed, suggesting strongly that biological reactions are stereochemical and occur on three-dimensional templates. It is also common in synthetic polymers (e.g. polypropylene, PVC, PMMA, polystyrene), where each alternate carbon atom in the chain has different substituents. This is known as 'tacticity' and it leads to polymers with markedly different properties.

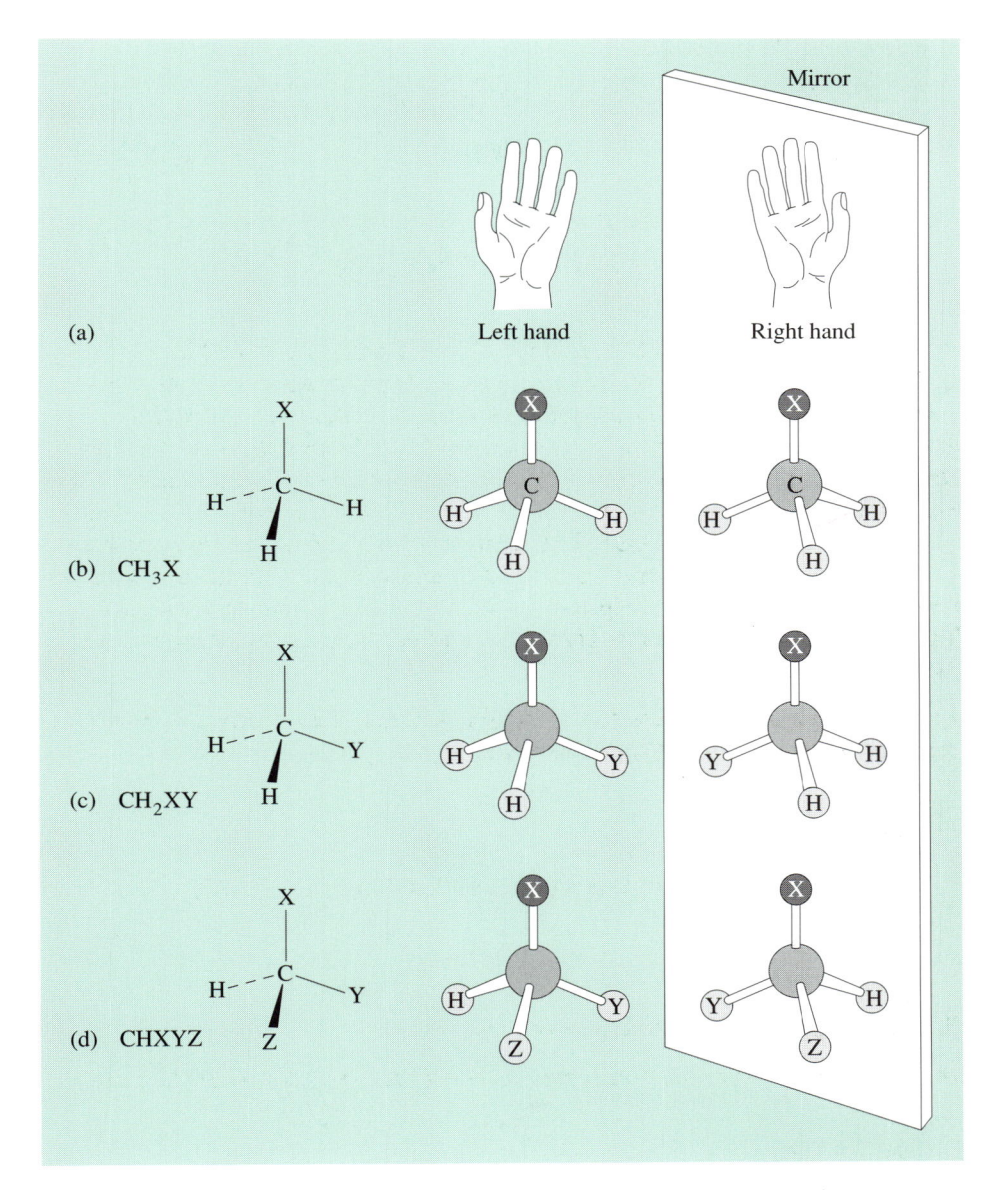

Figure 5.21 Mirror image symmetry at a single carbon atom with various substituents.

Returning to the linear, saturated hydrocarbons, there are two further points to be made. The first is that associated with the regular addition of each CH_2 group there is a regular shift in properties. This is exemplified in Table 5.14 by density, melting temperature and boiling temperature, which all increase monotonically with increasing molecular mass. Such behaviour is typical of all such series of molecules. These are known as homologous series, a hallmark of which is that they can be represented by a general formula. That for the series in Table 5.14 is C_nH_{2n+2}. (Note, however, that the densities are measured at different temperatures.)

> **Exercise 5.13** Fill in the final column of Table 5.14, selecting from 'waxy solid', 'liquid', 'solid polymer' and 'gas'.

Table 5.14 Densities and transition temperatures of linear saturated hydrocarbons

Name	Formula	Density of liquid (l) or solid (s) ρ/kg m^{-3}	T_m/K	T_b/K	State at 20 °C
ethane	CH$_3$.CH$_3$	509 (l)	89.9	185	
butane	CH$_3$.CH$_2$.CH$_2$.CH$_3$	579 (l)	135	273	
hexane	CH$_3$.(CH$_2$)$_4$.CH$_3$	660 (l)	178	342	
octane	CH$_3$.(CH$_2$)$_6$.CH$_3$	703 (l)	216	399	
decane	CH$_3$.(CH$_2$)$_8$.CH$_3$	730 (l)	243	447	
dodecane	CH$_3$.(CH$_2$)$_{10}$.CH$_3$	749 (l)	264	489	
tetradecane	CH$_3$.(CH$_2$)$_{12}$.CH$_3$	763 (l)	279	527	
hexadecane	CH$_3$.(CH$_2$)$_{14}$.CH$_3$	773 (l)	291	560	
octadecane	CH$_3$.(CH$_2$)$_{16}$.CH$_3$	777 (l)	301	589	
dotriacontane	CH$_3$.(CH$_2$)$_{30}$.CH$_3$	812 (s)	343	740	
heptacontane	CH$_3$.(CH$_2$)$_{68}$.CH$_3$	825 (s)	378	—	
polyethylene	CH$_3$.(CH$_2$)$_n$.CH$_3$ ($n \approx$ several hundred)	~850 (s)	400	decomposes	

The second point is that linear molecules are not the only possible structures. For instance, you can have branched structures. If you take propane (C$_3$H$_8$) and substitute a methyl (CH$_3$) group for one of the hydrogens on the central carbon atom, you get:

```
         H
         |
 H₃C— C —CH₃
         |
        CH₃
```

This is an **isomer** (i.e. a substance with the same molecular formula) of butane (C$_4$H$_{10}$), but can also be regarded as 2-methylpropane (the 2 indicating that the substituent is on the second carbon atom), or even as trimethylmethane. So, which is it? There is a vast set of internationally agreed rules governing the systematic naming of organic compounds, which is sometimes honoured more in the breach than the observance. This is particularly so for substances that have well-established common names, such as acetic acid, acetylene and formaldehyde that you met earlier (I also gave you their systematic names), and even polyethylene, which, under the rules should be polyethene. Luckily for you, the details of these rules are beyond the scope of this course, but they require that the molecule be called 2-methylpropane, since the longest straight chain contains three carbon atoms (confusingly, perhaps, they also allow it to be called *iso*butane). You'll be meeting both types of name in T203, reflecting current practice in the materials world.

Branching can have a significant effect on properties. This is particularly so for the long chain molecules of polymers. In the case of polyethylene, one process for making it produces straight chains, whilst another, earlier process produces branched chains. The straight chains can pack together efficiently to form crystalline regions, but the branches interfere with crystallization in the second type. The crystalline regions have a higher density and are stiffer than the amorphous regions, so the result is that the straight chain material has a higher density and higher elastic modulus than that containing branched chains. They are known as high-density polyethylene (HDPE) and low-density polyethylene (LDPE) respectively.

A further, important class of molecular structure is that of cyclic molecules in which the atoms are joined together in rings, which, in turn, can be joined to other rings or to noncyclic groups. You already saw some of these in the structure of buckminsterfullerene in 'Carbon'. Rings of carbon atoms can be formed with practically any number of atoms from three upwards, but perhaps the most important are those containing six atoms. There are two possible forms: cyclohexane, C_6H_{12}, and benzene, C_6H_6 (Figure 5.22).

In cyclohexane, all the carbon–carbon bonds are single, and the ring is puckered in a similar way to that in the linear hydrocarbons. Benzene, on the other hand, is a planar ring. In total its bonds are the equivalent of three single plus three double bonds, but the double bonds are effectively smeared out over the whole ring. That's why you'll frequently find it represented by the symbol ⌬. Benzene and chemicals based on it are extremely important industrially. As a substituent, known as the phenyl group, it forms part of polystyrene, many epoxy resins and phenolic resins such as Bakelite. It is also incorporated into the main chains of such polymers as polycarbonate, poly(ethylene terephthalate) or PET and the aramid fibres such as Kevlar.

Finally on covalent bonding, owing to the high bond strengths, many anionic groups are formed by different kinds of atom bonding covalently. These result in ionic compounds with more complex structures than you saw in the previous section. Often these also involve metal atoms sharing electrons. Some of the more common of these anionic groups are carbonates, nitrates and sulphates. Note also that the hydroxyl ion (OH^-) forms compounds called hydroxides.

SCIENCE FOR MATERIALS

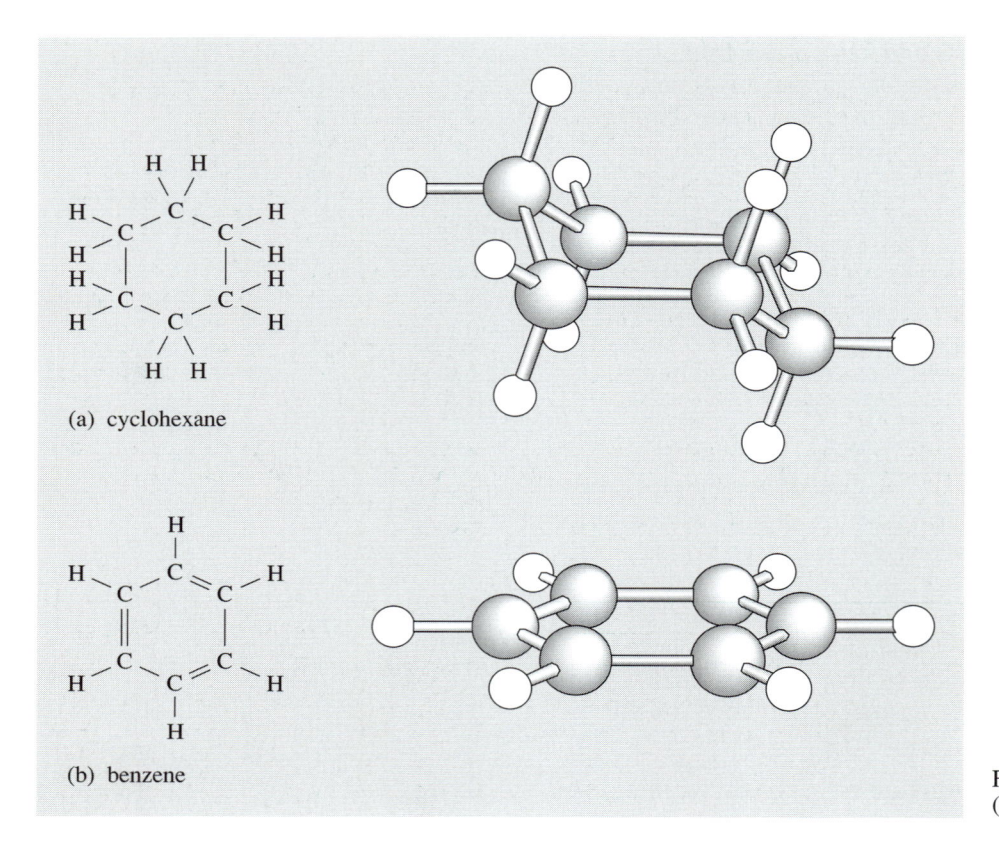

Figure 5.22 Structures of (a) cyclohexane, (b) benzene

5.5.3 Chemical periodicity and trends

You've already seen that there's a clear correlation between an element's position in the Periodic Table and its valency, and that valency determines the composition of the compounds that an element forms. This is the basis on which Mendeleev worked (see ▼Mendeleev and the Periodic Table▲). A full version of the modern Periodic Table, including electronic structures and the transition metals, is shown at the back of this book. You'll notice that the transition series have also been allocated groups, designated 1b to 7b (but not in that order) and then 8, and the original groups are labelled 1a to 7a followed by 0. I'll come back to this later.

Figure 5.23 shows the main named groups of elements in the Periodic Table. Many of these should already be familiar. The general trend is increasingly metallic from top to bottom of the Table and from right to left of the Table. The combination of both these trends means that a relatively small section of non-metals is created in the top right-hand corner of the Table. You'll also notice that there's an intermediate group of six elements known as semi-metals (or semiconductors) sandwiched between the metals and non-metals. Their electrical conductivity can be controlled very precisely from insulator to conductor by small additions of dopant atoms.

▼Mendeleev and the Periodic Table▲

You've already encountered the Russian chemist Dmitri Mendeleev (1834–1907) in the audiovisual exercise, so you should remember that he developed the very first Periodic Table of the elements in 1869. He was working on the preparation of a new textbook of chemistry, and wrote out a set of some 60 cards, one for each known element, listing their atomic masses and the formulae for the chemical compounds they formed with hydrogen and with oxygen. By moving these cards around (apparently, he liked to relax by playing patience), he sorted the elements into groups of those that formed similar compounds and he then arranged the cards in each group vertically with atomic masses increasing from top to bottom.

Next, he placed these groups alongside one another so that atomic masses increased from left to right. The result is shown in Table 5.15, where R is a general symbol for any element and the relative atomic masses are shown in brackets.

You can see that this is somewhat different from the Periodic Table in Frame 8 in Section 5.3. For example, there are no noble gases, there are twelve periods rather than seven and there are many gaps (dashes) and question marks in the table. Four of the gaps are highlighted and contain Mendeleev's predicted values of atomic mass. The first gap occurs in Group 3 in the fourth period for an element with a conjectured atomic mass of 44. This turned out to be 'Scandium', discovered in 1876 whose correct atomic mass is 45.0 (Table 5.5). The next two are 'Gallium' (relative atomic mass 69.7 rather than 68) and germanium (relative atomic mass 72.6, which is pretty close to 72) discovered in 1875 and 1886 respectively. The final one corresponds to technetium (relative atomic mass 98.9) which I said earlier has no stable isotopes, and has not been found naturally on Earth. This wasn't discovered until 1937.

Mendeleev's systematic approach was thus extremely valuable in enabling the existence of unknown elements to be predicted.

Table 5.15 Mendeleev's Periodic Table

Period	Group 1 RH R_2O	Group 2 RH_2 RO	Group 3 RH_3 R_2O_3	Group 4 RH_4 RO_2	Group 5 RH_3 R_2O_5	Group 6 RH_2 RO_3	Group 7 RH R_2O_7	Group 8 -- RO_4
1	H (1)							
2	Li (7)	Be (9.4)	B (11)	C (12)	N (14)	O (16)	F (19)	
3	Na (23)	Mg (24)	Al (27.3)	Si (28)	P (31)	S (32)	Cl (35.5)	
4	K (39)	Ca (40)	44	Ti (48)	V (51)	Cr (52)	Mn (55)	Fe (56), Co (59), Ni (59), Cu (63)
5	[Cu (63)]	Zn (65)	68	72	As (75)	Se (78)	Br (80)	
6	Rb (85)	Sr (87)	?Y (88)	Zr (90)	Nb (94)	Mo (96)	100	Ru (104), Rh (104), Pd (106), Ag (108)
7	[Ag (108)]	Cd (112)	In (113)	Sn (118)	Sb (122)	Te (125)?	I (127)	
8	Cs (133)	Ba (137)	?Dy (138)	?Ce (140)				
9								
10			?Er (178)	?La (180)	Ta (182)	W (184)		Os (195), Ir (197), Pt (198), Au (199)
11	[Au (199)]	Hg (200)	Tl (204)	Pb (207)	Bi (208)			
12				Th (231)		U (240)		

SCIENCE FOR MATERIALS

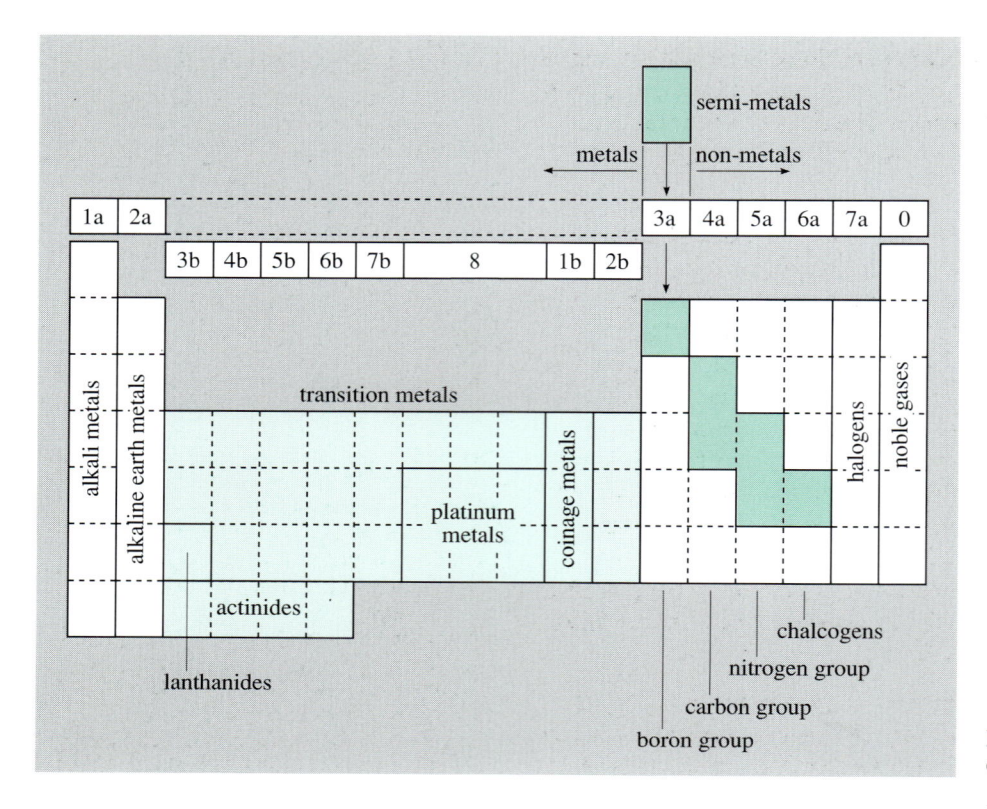

Figure 5.23 Named groupings of elements in the Periodic Table

Group 4a (Figure 5.24) contains five technologically important materials which provide a good example of this trend towards increasingly metallic nature. It starts with carbon, which is a non-metal. This is followed by silicon and ▼Germanium▲, which are both semiconductors, and then two metals, ▼Tin▲ and ▼Lead▲.

Members of any one of the Table's groups, particularly those which contain only metals or only non-metals, tend to have have similar but *not* identical properties. There are also clear trends in behaviour within such groups. For example, in some groups, elements become increasingly reactive with *increasing* atomic number. This happens with the alkali metals: lithium reacts only moderately with liquid water, potassium and sodium violently, rubidium very violently, but caesium reacts explosively. The main reason for this is the increasing size of the electron cloud moving down a group. The larger the cloud, the less strongly held the outer electrons are, and so they are more easily transferred in chemical reactions.

A similar trend is observed with the alkaline earth metals in Group 2, but when you get to the halogens in Group 7, this trend is reversed. Thus fluorine is an extremely reactive gas, chlorine a less reactive gas, bromine a liquid and iodine a relatively stable solid at ambient temperatures. It was established relatively recently that the Group 1 trend reasserts itself with the noble gases. Although helium, argon and neon do not form any compounds, xenon and krypton will form stable compounds when reacted with fluorine and oxygen.

6 C —$2s^2 2p^2$
14 Si —$3s^2 3p^2$
32 Ge —$4s^2 4p^2$
50 Sn —$5s^2 5p^2$
82 Pb —$6s^2 6p^2$

Figure 5.24 Group 4a of the Periodic Table

A striking feature of the Periodic Table is the large block of transition metals between Groups 2 and 3. As you saw, this block is expanded by the fourteen elements of the lanthanides, and there's another block — the actinides — with four naturally occurring members plus other, radioactive ones synthesized in nuclear reactors.

The properties of the transition metals are anomalous. For example, the group trend toward increasing reactivity *down* a group in Groups 1 and 2 is reversed within any small vertical group in this block of elements. This trend is immediately obvious in the coinage metals group: Cu, Ag and Au. Although copper is a reasonably stable metal, it will react with certain acids (nitric acid for example) to form salts rather easily. Silver is much less reactive, although silver spoons will tarnish in contact with sulphur-containing substances (such as egg white) to form a black coating of silver sulphide. By contrast, gold is inert to almost all reactive chemicals: the metal won't tarnish in either air or liquids. Neighbouring groups show similar behaviour. The 'platinum' metals (Ru, Rh, Pd, Os, Ir and Pt), rather like gold, are all rare and found in ores associated with one another. They're also inert and unreactive (except under special circumstances, when they catalyse other reactions, but themselves remain unaffected).

Most transition metals, especially the commercially important metals of the first series (scandium to zinc, see Table 5.8 on p. 37), show multiple valency as well as forming stable covalent bonds with non-metallic compounds. Thus iron forms Fe^{2+} and Fe^{3+} ions almost equally easily, and its covalent complex with a large organic molecule, haemoglobin, is literally vital to us as the oxygen-carrying compound in blood. There are some similarities between the various subgroups in the transition series and the main block elements: the coinage metals tend to be monovalent, whilst zinc, cadmium and mercury are divalent, for example. Such properties have led to the classification of the subgroups denoted 'b', as opposed to 'a' for the main groups (e.g. the coinage metals are designated 1b), as indicated in Figure 5.23 and in the main Periodic Table.

> **SAQ 5.9** (Objective 5.7)
> Indicate the physical and chemical properties you would expect of the following elements:
> (a) boron
> (b) selenium
> (c) magnesium
> (d) titanium.
>
> Use group and period trends as well as transition temperature data in your assessment. Include any information you might have on individual engineering applications.

In the next Unit, you'll look at chemical reactions and see how the thermodynamics of Unit 4 can be applied to their energetics.

▼Germanium▲

The second most important semiconductor material after 'Silicon', beneath which it sits in the Periodic Table, germanium (symbol Ge, atomic number 32, relative atomic mass 72.6, density 5320 kg m^{-3}) forms grey-white, brittle crystals. Apart from its use in electronic components, it's also used as a phosphor in fluorescent lamps, as a catalyst, as an ingredient in special optical glasses, and in alloying.

▼Tin▲

One of the elements that exhibits 'Allotropy', the normal form of tin (symbol Sn, atomic number 50, relative atomic mass 119, density 7290 kg m^{-3}) is β-tin or white tin. Below 13.2 °C it changes to the more brittle α- (or grey) tin (density 5770 kg m^{-3}), whilst above 161 °C it changes to γ- (or brittle) tin. These changes tend to be slow, and can be hindered by alloying. The major use of tin is in tin-plate (tin-coated steel sheet) used for making 'tin' cans. It's also a constituent of many alloys, e.g. solder, bronze, phosphor bronze, pewter. The float glass process for making flat glass relies on pouring the molten glass onto a bath of molten tin, on which it floats as it cools and solidifies.

▼Lead▲

Known certainly since Biblical times ('They sank as lead in the mighty waters' Exodus XV, 10), lead (symbol Pb, atomic number 82, relative atomic mass 207, density 11350 kg m^{-3}) is a very malleable, soft, corrosion-resistant metal. Despite being a cumulative poison, it was used from Roman times until this century for water pipes and other plumbing. Its major uses are in lead–acid batteries, ammunition, cable sheathing, X-ray and nuclear radiation shielding and as a sound and vibration absorber. Its alloys include soft solder, various bearing metals, pewter and metals for printing type. Lead glass is glass containing from 3–4% up to 50% by weight of lead oxide which is used for decorative glassware and for windows into radioactive areas.

Summary

- The drive to form chemical bonds is to achieve greater stability (i.e. the noble gas structure) with a consequent overall reduction in energy.
- The three types of primary bond are the ionic, covalent and metallic bonds: two types of secondary bond are van der Waals and hydrogen bonds.
- Ionic bonding involves the transfer of one or more electrons from one atom to another to form positively charged cations and negatively charged anions. The numbers of electrons transferred and the types of ions formed are governed by the valencies of the atoms involved. Ionic bonds are electrostatic and non-directional. Ionic molecules don't exist — ionic solids consist of arrays of cations and anions (a superlattice) arranged so that a balance is achieved between electrostatic repulsion and attraction.
- Covalent bonding involves the sharing between atoms of electrons in molecular orbitals. These can form single (two electrons), double (four electrons) or triple (six electrons) bonds, which are directed in space, and which produce discrete molecules. This can occur between atoms of the same element or between dissimilar atoms. An important example is the carbon–carbon bond, which can lead to linear or branched chains, two-dimensional networks (including rings) and three-dimensional arrays.
- Metallic bonding involves the sharing of a 'sea' of electrons amongst the atoms of the whole solid. It is a non-directional bond and has associated with it the high electrical and thermal conductivities characteristic of metals.
- The 'periodic' trends in physical properties of the elements are matched by similar trends in chemical behaviour.

Objectives for Unit 5

Having studied this Unit, you should be able to do the following.

5.1 Perform simple calculations using A_r to determine masses of elements present in substances of given chemical formulae (SAQs 5.1 and 5.2).

5.2 Explain the basis of the Periodic Table of the elements, describe its principal features and identify some of its main groupings (SAQ 5.3).

5.3 Determine electronic composition and configuration of sub-shells, shells and elements of given atomic number (SAQs 5.4, 5.5 and 5.8)

5.4 Account for the structure of the Periodic Table in terms of electronic configuration (SAQ 5.6).

5.5 Recognise and describe the main features of covalent, ionic, metallic, hydrogen and van der Waals bonding (SAQ 5.7).

5.6 Determine the valencies and types of ions formed by different elements (SAQ 5.8).

5.7 Predict the properties of elements from a knowledge of group and period trends in the Periodic Table (SAQ 5.9).

5.8 Define, describe or otherwise explain the meaning of the following terms:

- alkali metals
- alkaline earth metals
- allotropy
- anion
- asymmetric atom
- atomic number
- azimuthal quantum number
- cation
- CCP
- close-packed
- coordination number
- covalent bonds
- cubic close-packed
- electron volt
- empirical formula
- functional groups
- group
- halogens
- HCP
- hexagonal close-packed
- hydrocarbons
- hydroxonium ion
- ionic bond
- ionization energy
- ions
- isomer
- isotopes
- lanthanides
- magnetic quantum number
- molecular formula
- molecular orbitals
- noble gases
- orbital
- Pauli exclusion principle
- period
- Planck's constant
- primary bonds
- principal quantum number
- quanta
- quantum
- quantum numbers
- relative atomic mass
- saturated
- secondary bonds
- sp^3 hybrid
- spectrum
- spin quantum number
- stereoisomer
- structural formula
- superlattice
- transition metals
- unsaturated
- valency

Answers to exercises

EXERCISE 5.1 An atom is the smallest unit of a chemical element that retains the identity of that element. An atom comprises a positively charged nucleus surrounded by negatively charged electrons, so that the whole is electrically neutral (Figure 5.25).

The nucleus contains protons, each with one unit of positive charge, and electrically neutral neutrons, which bind the nucleus. It is only about 10^{-12} of the volume of the atom, yet represents virtually all of its mass, since the protons and neutrons are each about 2000 times more massive (actually 1840 ×) than the electrons. The electrons each carry unit negative charge, so must be equal in number to the protons for the atom to be electrically neutral. They can be visualized as occupying orbits around the nucleus at fixed levels of energy.

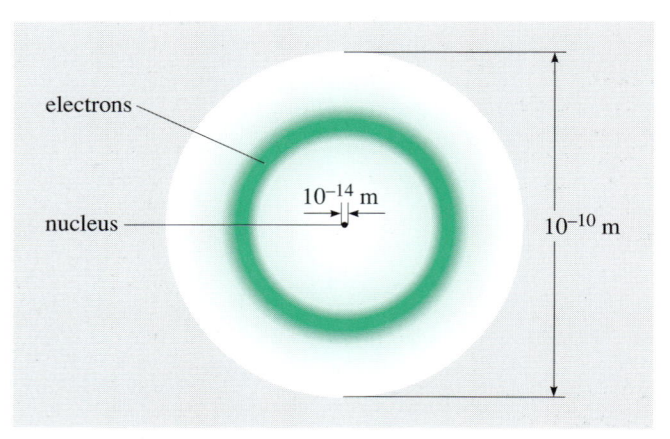

Figure 5.25 Representation of an atom

EXERCISE 5.2 A 1 mm cube has a volume of $(10^{-3})^3 \, \text{m}^3 = 10^{-9} \, \text{m}^3$. You should recall from earlier (e.g. 'Areas and volumes' on p.30 of Unit 1) that the volume of a sphere is $\frac{4}{3}\pi r^3$, where r is its radius. So the volume of an atom is about $\frac{4}{3}\pi(5 \times 10^{-11})^3 \, \text{m}^3$. Thus the number of atoms can be estimated from

$$\frac{\text{volume of cube}}{\text{volume of one atom}}$$

$$= \frac{3 \times 10^{-9}}{4\pi(5 \times 10^{-11})^3}$$

$$= 1.9 \times 10^{21} \text{ atoms.}$$

EXERCISE 5.3 Elements are substances that cannot be subdivided further by chemical means. They are chemically distinct types of atom. Each element has a characteristic number of protons in its nucleus, and it's this number, the **atomic number**, which is used to distinguish one element from another. It follows that an electrically neutral atom has the atomic number of electrons. In any one element, the number of neutrons is variable (though one number is usually dominant), giving rise to **isotopes** of the same element with the same atomic number, but with different mass.

EXERCISE 5.4 Relative atomic mass is the mass of an atom expressed on a scale of atomic mass units in which the carbon atom is defined to have a value of exactly 12. (1 atomic mass unit ≈ mass of a proton = 1.67×10^{-27} kg.) Atomic masses are not in general whole numbers because most elements are mixtures of isotopes, and what A_r represents is the mass of an average atom of an element taking into account the masses and relative abundances of its isotopes.

EXERCISE 5.5 From Table 5.5, the atomic number of argon is 18, so there are 18 protons, and these correspond to 18 units of atomic mass (the mass of the electrons is negligible). It has $A_r = 40.0$, and neutrons have almost the same mass as protons, so there must $(40 - 18) = 22$ neutrons.

EXERCISE 5.6 The N-shell has $n = 4$, so that l will have integral values from 0 to $(n - 1) = 3$. It can therefore have values of $l = 0, 1, 2$ and 3.

EXERCISE 5.7 Irrespective of the value of n, the number of orbitals is determined solely by the azimuthal quantum number, l. For the f sub-shell, $l = 3$, so the number of orbitals is $(2l + 1) = 7$. The corresponding magnetic quantum numbers, m, are +3, +2, +1, 0, −1, −2 and −3.

EXERCISE 5.8 The atom has the full orbital configuration of $1s^2 \, 2s^2 \, 2p^6 \, 3s^2 \, 3p^5$. So it has 2 electrons in the 1s orbital, 2 in the 2s orbital, 6 in the 2p orbital, 2 in the 3s orbital (all of these are paired), and 5 in the 3p orbital, one of which is unpaired. The total number of electrons is therefore 17.

EXERCISE 5.9 Compounds are formed as the result of chemical reactions and, in general, consist of elements combined (chemically bonded) in fixed proportions (e.g. water, H_2O). Mixtures are physical combinations of components whose proportions are not fixed and which can be separated by physical means (e.g. air, sea water). The properties of compounds are usually very different from those of their components (e.g. sodium chloride, NaCl, is very different from either sodium or chlorine), whilst those of mixtures tend to be intermediate between those of their constituents (e.g. air compared with oxygen and nitrogen).

EXERCISE 5.10
(a) You need two Ag^+ ions to neutralize O^{2+}, so the empirical formula must be Ag_2O.

(b) You need two Cl^- ions to neutralize Ca^{2+}, so the empirical formula must be $CaCl_2$.

(c) You need two Al^{3+} ions to neutralize three O^{2-}, so the empirical formula must be Al_2O_3.

EXERCISE 5.11 You saw in Table 5.12 that all these elements can accept a single electron to form anions and are therefore of single valency. Thus they should form single bonds between the pairs of identical atoms in diatomic molecules. The structural formulae are therefore

H—H (H_2)
F—F (F_2)
Cl—Cl (Cl_2)

EXERCISE 5.12 The simplest hydrocarbon must consist of a single carbon atom with hydrogen atoms attached to it:

```
     H
     |
H — C — H    or CH₄
     |
     H
```

This is methane, the principal component of natural gas.

EXERCISE 5.13

Name	State at 20 °C
ethane	gas
butane	gas
n-hexane	liquid
n-octane	liquid
n-decane	liquid
n-dodecane	liquid
n-tetradecane	liquid
n-hexadecane	liquid
n-octadecane	waxy solid
n-dotriacontane	waxy solid
n-heptacontane	waxy solid
polyethylene	solid polymer

AUDIOVISUAL FRAME 2 (c)
The noble gases are
He: helium
Ne: neon
Ar: argon
Kr: krypton
Xe: xenon.

AUDIOVISUAL FRAME 3

$T_b \approx 2 \times T_m$

AUDIOVISUAL FRAME 4(b)
The halogens are
F: fluorine
Cl: chlorine
Br: bromine
I: iodine.

AUDIOVISUAL FRAME 4(c)
The alkali metals are
Li: lithium
Na: sodium
K: potassium
Rb: rubidium
Cs: caesium.

AUDIOVISUAL FRAME 6

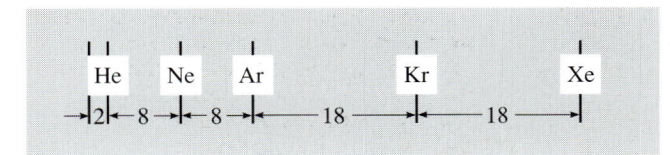

He	Ne	Ar	Kr	Xe
2	8	8	18	18

AUDIOVISUAL FRAME 7

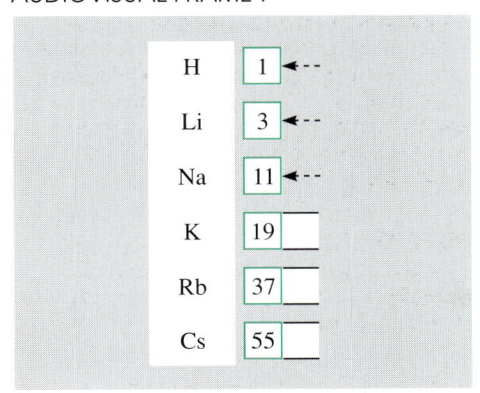

H	1
Li	3
Na	11
K	19
Rb	37
Cs	55

AUDIOVISUAL FRAME 8 The first series of transition metals

21 Sc	22 Ti	23 V	24 Cr	25 Mn	26 Fe	27 Co	28 Ni	29 Cu	30 Zn
scandium	titanium	vanadium	chromium	manganese	iron	cobalt	nickel	copper	zinc

Answers to self-assessment questions

SAQ 5.1 The chemical formula for water (H_2O) shows that there are two atoms of hydrogen present for every atom of oxygen. Their relative atomic masses (Table 5.5) are 1.01 and 16.0, so the ratio of the mass of oxygen to that of hydrogen in water is

$$\frac{16.0}{2 \times 1.01} = 7.92$$

If you examine their abundances in Table 5.1, you see that, for the hydrosphere the ratio is

$$\frac{1.5 \times 10^{21}}{1.8 \times 10^{20}} = 8.3$$

whilst that for the lithosphere is

$$\frac{1.1 \times 10^{22}}{2.4 \times 10^{19}} = 460$$

Clearly the ratio for the hydrosphere is very much closer to that for water than the one for the lithosphere.

SAQ 5.2 Simple inspection of the three ores shows that the limonite can be eliminated: it contains water (two H_2O molecules per Fe_2O_3), so will be less rich in iron than haematite (Fe_2O_3). The choice between haematite and magnetite (Fe_3O_4) can also be resolved easily: the atomic proportion of iron is greater in magnetite (three atoms out of the total of seven; 3/7 = 0.43) than in haematite (two out of five; 2/5 = 0.40). Since there are relatively more iron atoms in magnetite, it must be the optimum ore for the steelmaker. Now you need to calculate the mass composition:

mass % Fe

$$= \frac{3 \times A_r(Fe)}{3 \times A_r(Fe) + 4 \times A_r(O)} \times 100\%$$

$$= \frac{3 \times 56 \times 100}{(3 \times 56) + (4 \times 16)} \%$$

$$= \frac{168 \times 100}{168 + 64} \%$$

$$= \frac{168 \times 100}{232} \%$$

$$= 72.4\%$$

Thus, from 1000 tonnes of magnetite, the steelmaker should ideally be able to extract 724 tonnes of iron.

SAQ 5.3 The Periodic Table of the **elements** is based on systematic variations in physical and chemical properties with increasing **atomic number**. Amongst these properties is the **ionization energy**, which is the energy required to remove the **outermost** electron from an atom. Because the energies involved at an atomic scale are very **small**, these are often measured in **electron volts**. The noble gases have **high** ionization energies, reflecting their **low** chemical **reactivities**. The Table is arranged into **eight** vertical groups and **seven** horizontal periods. Groups 1 and 2 to the left of the Table contain the **alkali** and **alkaline earth** metals, respectively. On the right of the Table, Group 7 contains the **halogens** and Group **zero** the **noble gases**. Starting with the **fourth** period there is a gap between Groups **two** and **three** which is occupied by the **transition metals**.

SAQ 5.4 For any value of the principal quantum number, n, you first need to determine l, then m and finally s for each value of m. For $n = 5$ (O-shell), l can have values 0, 1, 2, 3 and 4 (s, p, d, f and g sub-shells).

For the s sub-shell ($l = 0$), m has $(2l + 1) = 1$ value and s can be $+\frac{1}{2}$ and $-\frac{1}{2}$. So only **2 electrons** are present in the s sub-shell.

For the p sub-shell ($l = 1$), m has $(2l + 1) = 3$ values (−1, 0, and +1), for each of which s can be + are $(3 \times 2) = $ **6 electrons** in the p sub-shell.

For the d sub-shell ($l = 2$), m has $(2l + 1) = 5$ values (−2, −1, 0, +1 and +2), for each of which s can be $+\frac{1}{2}$ or $-\frac{1}{2}$. So there are $(5 \times 2) = $ **10 electrons** in the d sub-shell.

For the f sub-shell ($l = 3$), m has $(2l + 1) = 7$ values (−3, −2, −1, 0, +1, +2 and +3), for each of which s can be $+\frac{1}{2}$ or $-\frac{1}{2}$. So there are $(7 \times 2) = $ **14 electrons** in the f sub-shell.

For the g sub-shell ($l = 4$), m has $(2l + 1) = 9$ values (−4, −3, −2, −1, 0, +1, +2, +3 and +4), for each of which s can be $+\frac{1}{2}$ or $-\frac{1}{2}$. So there are $(9 \times 2) = $ **18 electrons** in the f sub-shell.

The total number of electrons in the O-shell is thus (2 + 6 + 10 + 14 + 18) = 50. The total number of orbitals is just half this number (i.e. 25), since each orbital can only hold two electrons according to the Pauli exclusion principle.

SAQ 5.5
(a) Figure 5.26 shows Figure 5.8 with lines indicating the gaps of 2, 8, 8, 18 and 18 electrons. From what I said earlier, these lines are drawn just above the levels of the noble gases, and thus represent the positions of completed, full shells. You know (e.g. from Table 5.5) that krypton has $Z = 36$, so its outer shell is the first one containing 18 electrons, namely —$4s^2\, 3d^{10}\, 4p^6$.

(b) Using Figure 5.8 and the values of Z, you should have got Li —$2s^1$, Na —$3s^1$, K —$4s^1$ and Rb —$5s^1$. Each has one electron more than the noble gas that precedes it in atomic number, and this gives an outer shell comprising a single electron in an s-orbital. Each successive one has an increment in principal quantum number, n. They are, of course, the alkali metals, and make up Group 1 of the Periodic Table.

SAQ 5.6
(a) The configurations are:

noble gases: —ns^2np^6

alkali metals: —ns^1

alkaline earth metals: —ns^2

halogens: —ns^2np^5.

(b) The reason for the transition metals occupying the positions they do in the Periodic Table is that they represent the filling of d sub-shells. From principal quantum number $n = 3$ onwards, you have d-orbitals which can contain up to ten electrons. However, their energies are between those of the s-orbitals and those of the p-orbitals of the next shell up, so they occur in the period of the next higher principal quantum number, and fill after the s and before the p. Group 2 has a full s sub-shell, and Group 3 has the p-orbitals starting to fill, hence the ten transition metals between the two. From $n = 4$ onwards, you get f sub-shells, but their energies are between those of the $(n + 2)$ s sub-shell and the $(n + 1)$ d sub-shell. Thus the fourteen lanthanides come after Group 2 in the sixth period, and before the third series of transition metals.

SAQ 5.7 The three types of primary bonding are **ionic**, **covalent** and **metallic** bonds. The first of these involves the transfer of **electrons** from one type of atom to another and the bonding, which is **non-directional**, results from the

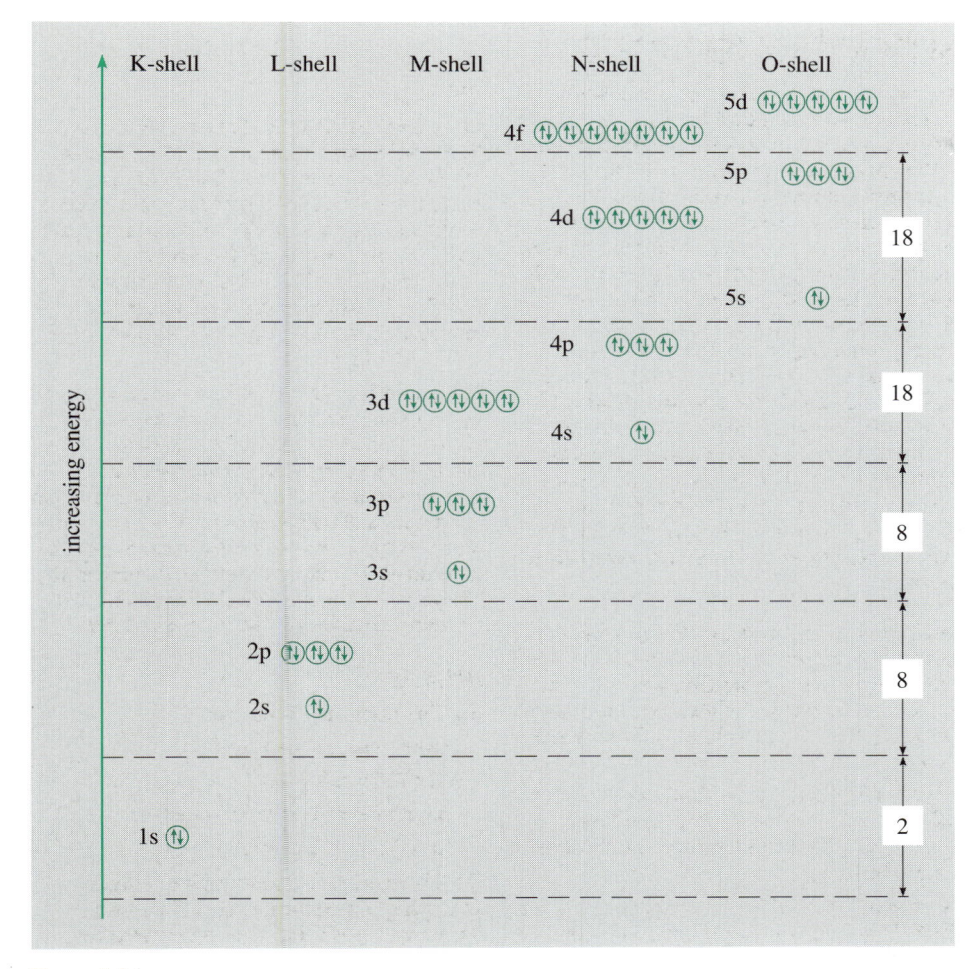

Figure 5.26

electrostatic attraction between oppositely charged **ions**. In the second and third, electrons are **shared** between atoms. In **covalently** bonded materials, the shared electrons form **directional** bonds between pairs of adjacent atoms. Such bonding can result in molecules with as few as **two** atoms (**diatomic molecules** such as hydrogen or **oxygen**) and up to the very large numbers of atoms in the giant molecules called **polymers**. The resulting structures can be **linear** molecules, two-dimensional **networks**, or **three-dimensional** arrays. The third type is non-directional, and applies only to **metals**, where the bonding electrons are **delocalized** and spread out over the array of atoms.

The two types of secondary bonding are **hydrogen** bonding and **van der Waals** bonding. The first results from an **uneven** charge distribution in molecules such as **water**, whilst the second results from **fluctuations** with **time** in the charge distributions in covalently bonded materials.

SAQ 5.8 Apart from helium ($1s^2$), the noble gases have —ns^2np^6. Elements in Group 2 are the alkaline earth metals which have —ns^2 as their outer electron configuration. The nearest noble gas has two electrons less, so they would need to *lose* two electrons. (They could also gain six electrons, but this is less favourable energetically.) They form divalent, positively charged cations (their valency is 2). Group 7 elements (the halogens) have —ns^2np^5, so they would only need to *gain* a single electron (as opposed to losing five). They thus form monovalent, negatively charged anions (i.e. their valency is 1).

SAQ 5.9
(a) With its small atom and three outermost electrons, boron can be expected to form trivalent compounds (e.g. B_2O_3 with oxygen) in a similar way to aluminium. Most of its compounds will be covalent since the element sits close to the non-metal block. Its elemental state may also be expected to be covalent ($T_m \approx 2600$ K from Figure A of Frame 2) and strongly bonded, like adjacent carbon (boron whiskers are used in composite materials). It should also be a semiconductor, showing some non-metallic properties. Thus, it should form borides (B^{3-}) with highly metallic elements (alkali, alkaline earths).

(b) Selenium should behave as a typical non-metal : non-conductor of electricity etc. Its valency will be two and it will show greatest similarity to sulphur in properties. For example, it should form selenides (like Li_2Se) very easily (just like sulphur). It possesses a low T_m of about 500 K, slightly greater than that of sulphur. However, it is close to the semi-metals and being lower in its Group (six) may show some unusual properties. In fact it's photoconductive and widely used for the photocopying or xerographic process.

(c) As an alkaline earth, magnesium has two outermost electrons which are easily lost to form a divalent cation (Mg^{2+}). Its reactivity will be relatively low, being near the top of Group 2, at least in the bulk state. Having a low atomic number, it will have a low density and should therefore be useful for light alloys (perhaps with aluminium, the light metal next to it in the third period). It's used, for example, in airframe structures. Its melting point is relatively low (1000 K) and almost identical with that of aluminium.

(d) Titanium is a member of Group 4b and the first transition series. Its valency will be 4, so will form an oxide of formula TiO_2. This is a white pigment, widely used in paints etc. It might be expected to have other valence states however (as a transition metal). Some covalent bonding may occur in its compounds. Its high T_m of about 1700 K will make it a useful alloy, especially for its relatively low density (atomic number 23, relative atomic mass 48).

Unit 6: Chemical reactions

Contents

6.1	Introduction	66
6.2	Thermochemistry	73
6.3	Equilibrium reactions	90
6.4	Reaction kinetics	105
Objectives for Unit 6		117
Answers to exercises		118
Answers to self-assessment questions		119

6.1 Introduction

In the last unit, you saw how the 92 elements could be classified in terms of their physical properties and chemical behaviour, and how this was a reflection of their electronic structure. You also saw that this structure influenced the types of chemical bond that they formed, both with themselves and with other elements. The drive to form such bonds was likened to the drive in mechanical systems to minimize potential energy. In Unit 4, this was generalized as the drive to minimize thermodynamic free energy in systems especially when thermal energy is involved.

In this Unit, you'll be addressing the following questions.
- Why does a reaction happen?
- If it does, what are the energy outputs or inputs?
- How far does the reaction go?
- How fast does it go?

But first, let's establish some of the terminology of reactions.

6.1.1 The nature of reactions

A chemical reaction is the general term for any process that involves the formation or breakdown of chemical bonds. Such processes are generally expressed by chemical equations, as ▼Reading chemical equations▲ reminds you. Their general form is simply

$$(\text{reactants}) \longrightarrow (\text{products})$$

Stated in this way, the implication is that *all* of the reactants are converted to products, and this is often the case. However, in principle, all chemical reactions can be reversed under the right conditions (e.g. temperature, pressure, concentration). Where the conditions for the reverse reaction are very similar to those for the forward reaction, both will occur at the same time. A point will be reached where the concentrations of reactants and products no longer change — a state of dynamic equilibrium has been achieved in which the rate of the forward reaction is just balanced by the rate of the reverse reaction. Such **equilibrium reactions** are denoted by a double arrow, i.e.

$$(\text{reactants}) \rightleftharpoons (\text{products})$$

A simple example of this is dissolving common salt (sodium chloride, NaCl) in water producing sodium ions (Na^+) and chloride ions (Cl^-). At any given temperature, only a finite amount of the salt will dissolve. At this limit, any further salt going into solution will cause other ions to precipitate out of solution to reform salt crystals. The only way to get more salt into solution is to alter the conditions, for example add more water (though this won't affect the final concentration) or increase the temperature (which will). Under given conditions, the reaction is expressed as

$$NaCl(s) \underset{}{\overset{H_2O}{\rightleftharpoons}} Na^+(aq) + Cl^-(aq)$$

▼Reading chemical equations▲

A **chemical equation** is a shorthand way of writing down a chemical reaction using the chemical symbols of the elements and compounds involved. Chemical equations have certain similarities with mathematical equations, but there are also important differences. At its simplest, a chemical equation has the form

(reactants) $\longrightarrow$ (products)

Note that an arrow ($\longrightarrow$) is used rather than the equals sign of a mathematical equation, and indicates the direction of the reaction. Many chemical reactions can go in either direction (**reversible reactions**), dependent on conditions. You can express this with a double arrow, i.e.

(reactants) $\rightleftharpoons$ (products)

(although if one direction predominates only a single arrow is often used). You can also write over the arrows to indicate the conditions under which a reaction occurs; e.g.

(reactants) $\xrightarrow{H_2O}$ (products)

shows that the reaction takes place in the presence of water, and

(reactants) $\xrightarrow{650\,K}$ (products)

shows that the reaction takes place at temperatures above 650 K.

Chemical equations also tell you the proportions in which the reactants combine and the proportions of the resulting products. Like algebraic equations, chemical equations have to be balanced, with elements taking the place of variables, so that the same amount of each participating element appears on each side of the equation. However, a somewhat different notation is used, as you've already seen a bit of in previous units. To summarize:

- A prefix integer tells you the number of moles of the atom, molecule or compound whose chemical formula immediately follows the integer, e.g. 2Fe, 3CH$_3$Cl, 4MgO, 5HCl.
- A subscript integer tells you the number of the immediately preceding atoms or groups that are chemically bound in a molecule or ionic compound, e.g. H$_2$O, Na$_2$SO$_4$, Fe(NO$_3$)$_2$, (NH$_3$)$_2$CO$_3$.

In addition, there can be:

- A superscript integer followed by a plus or minus sign which tells you the number and type of charge on the ion whose formula precedes it, e.g. Ca^{2+}, O^{2-}, Fe^{3+}, SO$_4^{2-}$, SiO$_4^{4-}$.
- The symbol 'e$^-$', which may be preceded by an integer, which tells you the number of electrons involved in an electrochemical reaction (e.g. electroplating, types of corrosion, batteries)
- Letters in brackets following the chemical formula of an atom, molecule or compound indicating its physical state — the ones you'll meet are (g) for gas, (l) for liquid, (s) for solid and (aq) for aqueous solution.

EXERCISE 6.1 Write down the names of the elements, compounds and ions given above.

Balancing an equation involves adjusting the prefix integers until there is the same total number of each type of atom on each side of the equation, and making sure that the set of integers represents the minimum values. (If you've ever tried doing this, you'll know that it isn't always as easy as it sounds.) Here's a simple example: you know that hydrogen and oxygen combine to form water, and that these elements exist as diatomic molecules. So you could write

$$H_2(g) + O_2(g) \longrightarrow H_2O(l)$$

This is unbalanced (count the 'O's), but if you double up on the hydrogen and the water molecules, you get

$$2H_2(g) + O_2(g) \longrightarrow 2H_2O(l)$$

which, with four 'H's and two 'O's on each side, is balanced.

Another example: the reaction between sulphuric acid (H$_2$SO$_4$) and sodium hydroxide (caustic soda, NaOH) produces sodium sulphate (Na$_2$SO$_4$) and water, or

$$H_2SO_4(aq) + NaOH(aq) \longrightarrow Na_2SO_4(aq) + H_2O(l)$$

The H, O, and Na can be made to balance by doubling the NaOH, giving

$$H_2SO_4(aq) + 2NaOH(aq) \longrightarrow Na_2SO_4(aq) + 2H_2O(l)$$

A final example: the equation for electroplating copper from copper sulphate (CuSO$_4$) solution could be written as

$$CuSO_4(aq) \longrightarrow Cu(s) + SO_4^{2-}(aq)$$

But this is unbalanced with respect to charge. You need two negative charges on the left-hand side to compensate for those on the sulphate SO$_4^{2-}$ ion, which is provided by putting two electrons (2e$^-$) in, to give

$$CuSO_4(aq) + 2e^- \longrightarrow Cu(s) + SO_4^{2-}(aq)$$

You'll probably be glad to learn that you won't be asked to balance many chemical equations in this course, but it's important for you to appreciate the need for it. There'll be an opportunity for you to practice the skill with some software at Summer School, should you want to.

What you are doing when you balance an equation is conserving mass — the total mass of products has to equal the total mass of reactants (note that this is not true for nuclear reactions, where mass and energy can be interconverted). But this doesn't mean that the equations are expressed in mass units. They represent the *numbers* of atoms, molecules or ions that combine with each other, and as you know, different atoms, etc. have different masses (remember relative atomic mass A_r, in Unit 5). Thus the equations can either be interpreted at the level of individual atoms, etc., or as an expression of the relative numbers of moles of each species involved.

Whilst you cannot have anything other than whole numbers of atoms or molecules in an equation, it's perfectly permissible to have fractional numbers of moles. Thus, the above equation for producing water could be written in molar terms as

$$H_2(g) + \tfrac{1}{2}O_2(g) \longrightarrow H_2O(l)$$

meaning that one mole of hydrogen combines with half a mole of oxygen to give one mole of water.

SAQ 6.1 (Objective 6.1)
Balance the following equations
(a) Mg(s) + O$_2$(g) $\longrightarrow$ MgO(s)
(b) HNO$_3$(aq) + Ca(OH)$_2$(aq) $\longrightarrow$ Ca(NO$_3$)$_2$(aq) + H$_2$O
(c) Fe(OH)$_2$(s) + O$_2$(g) + H$_2$O(l) $\longrightarrow$ Fe(OH)$_3$(s)
(d) Ag$_2$SO$_4$(aq) $\longrightarrow$ Ag(s) + SO$_4^{2-}$(aq)

You'll explore equilibrium reactions in more detail later. Whether a reaction goes to completion or not depends on the energy available. This is the realm of thermodynamics, so by 'energy' I mean 'free energy', namely ΔG for reactions at constant temperature and pressure (Unit 4). You'll see how thermodynamics can be applied to reactions in the next section. But there's another issue which also needs discussion — the *rate* at which reactions occur.

Irrespective of whether a given reaction goes to completion or only reaches equilibrium, there's the question of how fast it arrives at that position. Some reactions are extremely rapid, occurring in a time scale of microseconds (10^{-6} s) or milliseconds (10^{-3} s). Examples include explosions (Figure 6.1(a)), where such compounds as nitroglycerine react on detonation to produce large volumes of carbon dioxide, water vapour and nitrogen oxides. Other reactions take place in seconds or minutes, like many combustion reactions where a fuel burns in air or oxygen (Figure 6.1(b)). Yet others proceed so slowly that their progress can only be measured in months if not years. Degradation of engineering materials through corrosion (in the case of metals) or by weathering (in the case of polymers) are examples of such very slow reactions (Figure 6.1(c)). The rates (or kinetics) of reactions are also explored later in this Unit.

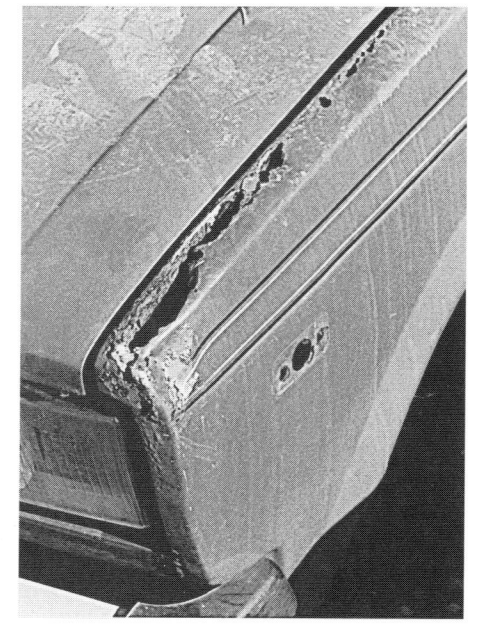

Figure 6.1 Reactions at different time scales: (a) an explosion; (b) combustion; (c) rusting of steel

Types of reaction

Despite the vast number of chemical compounds, the reactions involving them can be classified into a relatively small number of types. The ones you'll encounter include:

- oxidation, and its converse, reduction
- hydrolysis (i.e. what happens to compounds in water)
- reactions between acids and bases
- electrochemical reactions

Oxidation and reduction

At its simplest, **oxidation** is just the combining of oxygen with another element or group to form an oxide, for example with mercury

$$2Hg(s) + O_2(g) \longrightarrow 2HgO(s)$$

There is a very large number of such oxidation reactions, including combustion reactions (e.g. burning of fuels), pyrotechnic reactions (e.g. burning of metals for special effects, as in fireworks) and metabolic reactions (e.g. food conversion in living organisms). They are all important technologically for their role in producing energy, which might take the form of heat (fuels), light (pyrotechnics, fuels) or mechanical work (fuels, food for muscles).

The converse of oxidation is **reduction**, where an oxidized material is converted into an unoxidized form. Examples of such processes include the extraction of metals from ores (e.g. iron making), many petrochemical reactions (e.g. hydrogenation) and photosynthesis (in which plants reduce carbon dioxide to sugars and cellulose).

As the converse of oxidation, such processes usually require large inputs of energy, whether in the form of heat as in iron making, or light as in photosynthesis (Figure 6.2).

Many reduction reactions are just the reverse of simple oxidation reactions, e.g.

$$2HgO(s) \xrightarrow{710\,K} 2Hg(s) + O_2(g)$$

shows that this reverse reaction occurs at 710 K. But the term actually applies much more widely. For example, hydrogen gas can reduce many materials creating water as one of the products, e.g.

$$Fe_2O_3(s) + 3H_2(g) \xrightarrow{1300\,K} 2Fe(s) + 3H_2O(g)$$

This can be called a reduction reaction because the principal reactant, iron oxide, is converted to iron. However, the **reducing agent**, hydrogen is at the same time converted to water and is therefore oxidized. The generic term for such reactions in which one reactant reduces another and is itself oxidized in the process is a **redox reaction** (from reduction and oxidation).

(a)

(b)

Figure 6.2 Sites of two important reduction reactions: (a) a blast furnace where iron ore is reduced; (b) a leaf where light induces the reduction of CO_2

Hydrolysis

Since water (H_2O) is such a common material on Earth, its chemical interaction with various materials is often given a separate name, **hydrolysis**. You saw in Unit 5 that small amounts of hydroxyl ions (OH^-) and hydroxonium ions (H_3O^+) are normally present in neutral water. It is these very reactive ions which are the active components in hydrolysis, as with **alkalis** and **acids** respectively. For instance, a common alkali is sodium hydroxide (NaOH), which, when dissolved in water, splits up to produce sodium ions and hydroxyl ions:

$$NaOH(s) \xrightarrow{H_2O} Na^+(aq) + OH^-(aq)$$

The splitting-up of an ionic compound into its constituent ions is known as **dissociation**. Such alkalis are very reactive chemicals and are widely used for breaking down complex materials into simpler constituents by hydrolysis. Their reactivity comes from the increased concentration of hydroxyl ions.

Inorganic acids, such as hydrochloric acid (HCl) and sulphuric acid (H_2SO_4), are almost as reactive as alkalis. Their common characteristic is that hydrogen forms the cation (remember that a cation is a positive ion). On hydrolysis, their dissociations are

$$HCl(aq) \xrightarrow{H_2O} H^+(aq) + Cl^-(aq)$$

$$H_2SO_4(aq) \xrightarrow{H_2O} 2H^+(aq) + SO_4^{2-}(aq)$$

respectively, where, in each case, the protons (H^+) react further to give

$$H^+ + H_2O \longrightarrow H_3O^+$$

As you saw in Unit 5, this occurs because lone protons are so reactive. So with acids, it is the hydroxonium ions (H_3O^+) that are effectively the active species, so, for instance, the HCl reaction above could also be written

$$HCl(aq) + H_2O \longrightarrow H_3O^+ + Cl^-(aq)$$

Acids and bases

Early chemists knew of the existence of two classes of compounds, which they called acids and alkalis. Both classes were known to be sharp-tasting and usually corrosive. Acids were typified by such naturally occurring ones as vinegar. Alkalis were similar to the extracts from burnt seaweed ('alkali' is derived from the Arabic *al-qili*, the ashes of the saltwort), or from wood ash (this was burned in pots, hence 'potash'). To the early chemists, these two classes, despite their similarities, nonetheless represented two extremes of behaviour; for example, if an acid changed the colour of a vegetable dye, then an alkali tended to reverse the change. These extremes of behaviour were nowhere more noticeable than in the reaction between acids and alkalis themselves — when mixed together they neutralized each other.

Modern chemistry still recognizes these two classes of compound, although nowadays alkalis are placed in a larger class of compounds known collectively as **bases**. So, what happens with acids and bases if, instead of reacting with water, they react with each other? The reaction is general:

$$\text{acid} + \text{base} \longrightarrow \text{salt} + \text{water}$$

So, for example, potassium hydroxide reacts with sulphuric acid to give potassium sulphate (a salt) and water,

$$H_2SO_4(aq) + 2KOH(aq) \longrightarrow K_2SO_4(aq) + 2H_2O$$

In the organic world, things are slightly different, though analogous. The organic equivalent of an acid contains the carboxylic acid group ($-COOH$), and that of an alkali is an alcohol which contains the $-OH$ group. Because these are covalently bonded, the nature of the reaction differs from that of the ionically bonded inorganic acids and alkalis, but once again the reaction is general:

$$\text{organic acid} + \text{alcohol} \longrightarrow \text{ester} + \text{water}$$

An **ester** is the organic analogue of an inorganic salt, e.g. ethyl acetate as the product of the reaction between acetic acid and ethanol:

$$CH_3.COOH + C_2H_5.OH \longrightarrow CH_3.COO.C_2H_5 + H_2O$$

Both the inorganic and organic reactions produce water. Reactions in which two or more reactants combine so that one of the products is a simpler molecule or compound, e.g. water, are often known as **condensation reactions**.

Electrochemical reactions

Electrons are, of course, involved in all chemical reactions, since it is they that form and reform the bonds. Electrochemical reactions, such as occur in electrolytic cells (as used in electroplating and batteries) and in some types of corrosion, differ because there is a *flow* of charge involved, and this must be taken into account explicitly. You saw one example in 'Reading chemical equations' — that with copper sulphate ($CuSO_4$). Another one is described in ▼The lead–acid battery▲

SAQ 6.2 (Objective 6.2)
Classify the following reactions into reaction type

(a) $2CuO(s) \xrightarrow{1720\,K} 2Cu(s) + O_2(g)$

(b) $HNO_3 + H_2O \longrightarrow H_3O^+ + NO_3^-$

(c) $2Mg(s) + O_2(g) \longrightarrow 2MgO(s)$

(d) $C_2H_5.COOH + CH_3.OH \longrightarrow C_2H_5.COO.CH_3 + H_2O$

SCIENCE FOR MATERIALS

▼The lead–acid battery▲

Figure 6.3 shows schematically a lead–acid battery or accumulator (the type of battery fitted in cars and trucks). It has two sets of electrodes, one of plates of lead (Pb) paste in a lead grid, the other of a paste of lead dioxide (PbO_2) also held in a grid of lead. The electrodes are separated by a microporous insulating sheet, often the plastic PVC. The electrode assembly is immersed in sulphuric acid which provides H_3O^+ and SO_4^{2-} ions.

When the battery terminals are connected to an electrical circuit, a current flows due to the following reactions:

at the negative terminal (cathode)

$$Pb(s) + SO_4^{2-}(aq) \longrightarrow PbSO_4(aq) + 2e^-$$

at the positive terminal (anode)

$$PbO_2(s) + SO_4^{2-}(aq) + 4H_3O^+ + 2e^- \longrightarrow PbSO_4(aq) + 6H_2O$$

Notice that, as required, electrons are generated at the cathode and absorbed by the anode, giving a flow of charge round the circuit. Also that the reaction product is lead sulphate ($PbSO_4$) in both cases.

By adding these two reactions, you could write the net reaction as

$$Pb + PbO_2 + 2H_2SO_4 \longrightarrow 2PbSO_4 + 2H_2O$$

But you can see that this doesn't tell you as much about what's going on as the individual electrochemical reactions at each terminal.

As you probably know, one feature of batteries such as these is that they are rechargeable. By connecting them to a suitable source (e.g. a battery charger or alternator) and driving the electrons through them in the opposite direction, they can be restored to their original state of charge. This means that the above reactions must be reversible, so that they are more properly written as

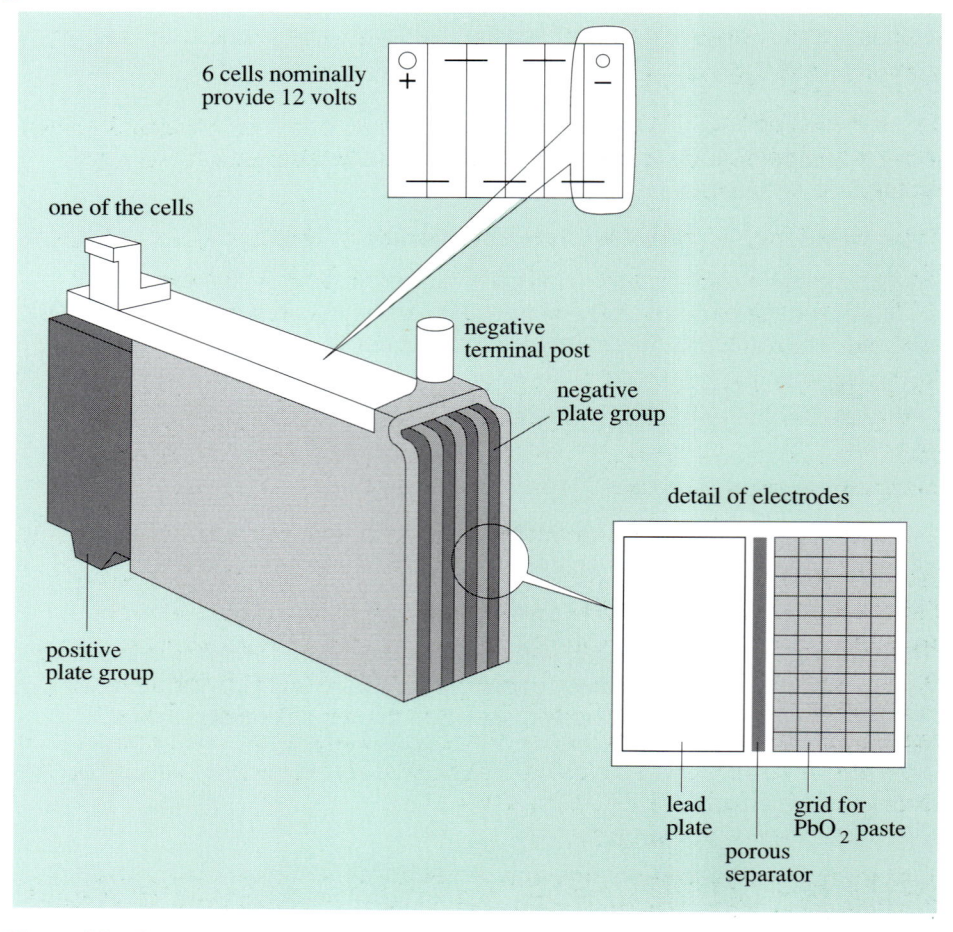

Figure 6.3 Construction of a typical 12 V lead–acid battery

$$Pb(s) + SO_4^{2-}(aq) \underset{\text{charge}}{\overset{\text{discharge}}{\rightleftharpoons}} PbSO_4(aq) + 2e^-$$

$$PbO_2(s) + SO_4^{2-}(aq) + 4H_3O^+ + 2e^- \underset{\text{charge}}{\overset{\text{discharge}}{\rightleftharpoons}} PbSO_4(aq) + 6H_2O$$

Thus the recharging process converts the lead sulphate back to lead at the cathode and to lead dioxide at the anode.

6.2 Thermochemistry

6.2.1 Some basics

The first half of Part 1 *Chemistry* of Videocassette 2 and its associated notes explore thermochemistry. You should study them in conjunction with this section.

You saw in Unit 4 that the thermodynamic criterion for a change to occur is that the change in free energy, ΔG is negative in the free energy equation (Equation 4.31)

$$\Delta G = \Delta H - T\Delta S \tag{6.1}$$

where ΔH is the enthalpy change and ΔS the entropy change at temperature T. You also saw that the thermal energy, Q, which is not a state function, could be replaced by the state function, enthalpy, for changes at constant temperature and pressure, i.e.

$$\Delta H = \Delta Q \text{ (constant } T, \text{ constant } p) \tag{6.2}$$

Most chemical reactions either produce or absorb heat (i.e. thermal energy). This results from the difference in bond energy between reactants and products and is shown schematically in Figure 6.4, where the thermal energy is expressed as enthalpy. Thus the change is the **reaction enthalpy change** (often abbreviated to 'reaction enthalpy') for the reaction in question. You can see that when thermal energy is released (ΔH negative on our sign convention) the reaction is called **exothermic**, and, conversely, when it is absorbed (ΔH positive) it's called **endothermic**.

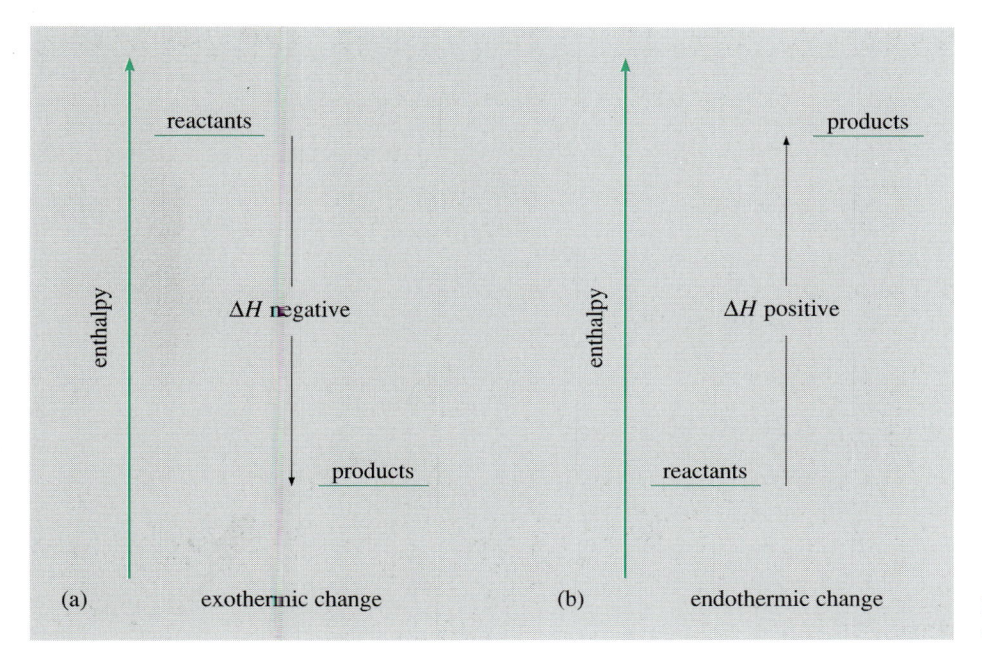

Figure 6.4 Schematic of enthalpy changes in chemical reactions

SCIENCE FOR MATERIALS

At constant temperature and pressure then, the heat input or output to an isolated system is a direct measure of the enthalpy change in the system. So how is it measured? The method is known as ▼Calorimetry▲ (notice the link with the old caloric theory of heat). The greatest source of information has come from relatively simple, easily controlled reactions such as combustion. If you look up published tables of thermochemical data you will often find tables of 'combustion enthalpies' (also called 'heats of combustion') and Table 6.1 lists typical examples. The reactants are initially at 298 K and the reaction products are either $H_2O(l)$ or $CO_2(g)$ or both, also at 298 K. Essentially these are the reaction enthalpy changes when the specified compound reacts completely with oxygen.

▼Calorimetry▲

Calorimetry measures changes in temperature in devices called calorimeters. For reaction calorimetry, these are essentially containers provided with heat exchangers. Water, for instance, is an ideal heat exchange medium. So surrounding the container with a jacket containing a known volume of water enables you to determine the total heat change in the reactor by measuring the temperature rise in the jacket (via the specific heat capacity of the water, c_p, and using $\Delta Q = mc_p\Delta T$).

Many calorimeters of different design have been constructed to measure reaction enthalpy changes for specific reactions. For example, the heat output of different fuels is of crucial important to industrial users since it determines (among other things) the efficiency of heat-controlled processes. Many processes for making materials, such as the making of iron in a blast furnace or the thermal cracking of naphtha to make ethylene monomer, are conducted at high temperatures. The heat content of solid fuels such as coal and coke as well as liquid fuels like oil can be determined in a bomb calorimeter (Figure 6.5).

The vessel is constructed of stainless steel to resist internal corrosion and has a capacity of 250–300 ml. It's strong enough to withstand pressures developed by the combustion of 1 g of fuel at an initial pressure of 30 atmospheres of oxygen gas — needed to ensure complete combustion of the sample. The calorimeter is immersed in a well-stirred water bath, itself lagged to maintain adiabatic conditions, and the temperature rise of the water is measured after the fuel has been ignited. From this value, the thermal energy given out by the fuel can be determined.

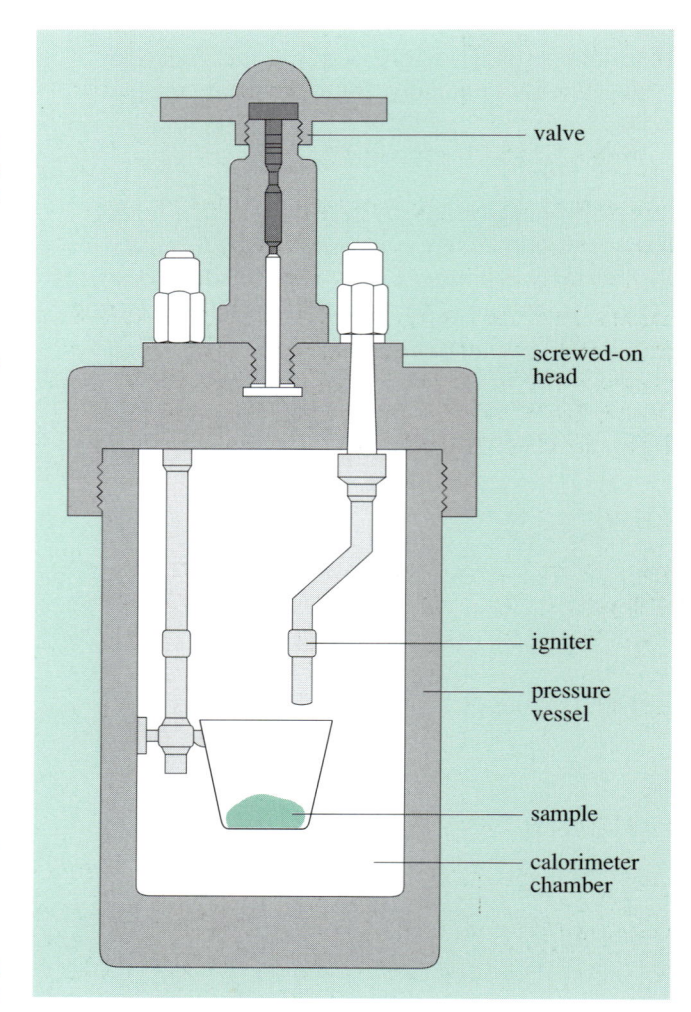

Figure 6.5 A bomb calorimeter

Table 6.1 Typical values of combustion enthalpy at constant pressure.

Reactant	Reaction	$\Delta H / \text{kJ mol}^{-1}$
hydrogen $H_2(g)$	$H_2(g) + \frac{1}{2}O_2(g) \longrightarrow H_2O(l)$	−286
graphite $C(s)$	$C(s) + O_2(g) \longrightarrow CO_2(g)$	−394
carbon monoxide $CO(g)$	$CO(g) + \frac{1}{2}O_2(g) \longrightarrow CO_2(g)$	−284
methane $CH_4(g)$	$CH_4(g) + 2O_2(g) \longrightarrow CO_2(g) + 2H_2O(l)$	−893
ethane $C_2H_6(g)$	$C_2H_6(g) + \frac{7}{2}O_2(g) \longrightarrow 2CO_2(g) + 3H_2O(l)$	−1566
pentane $C_5H_{12}(g)$	$C_5H_{12}(g) + 8O_2(g) \longrightarrow 5CO_2(g) + 6H_2O(l)$	−3550
octane $C_8H_{18}(g)$	$C_8H_{18}(g) + \frac{25}{2}O_2(g) \longrightarrow 8CO_2(g) + 9H_2O(l)$	−5492
ethylene (ethene) $C_2H_4(g)$	$C_2H_4(g) + 3O_2(g) \longrightarrow 2CO_2(g) + 2H_2O(l)$	−1416
benzene $C_6H_6(l)$	$C_6H_6(g) + \frac{15}{2}O_2(g) \longrightarrow 6CO_2(g) + 3H_2O(l)$	−3280

Usually combustion enthalpies are quoted as energy per mole of reactant, as in Table 6.1. It is, however, a simple matter to convert these to energies per unit mass (in kg). For example, the combustion enthalpy for pentane in Table 6.1 is given as −3550 kJ mol^{-1}. The relative atomic masses of carbon and hydrogen are 12 and 1 respectively, so the the relative molecular mass of pentane (C_5H_{12}) is:

$(5 \times 12) + (12 \times 1) = 72$

So 1 mole of pentane has a mass of 72 g or 0.072 kg. When this quantity is burned in oxygen, the enthalpy change is −3550 kJ. Thus, when 1 kg of pentane is burned completely in oxygen, the enthalpy change is

$\Delta H = -3550 \times \frac{1}{0.072} \text{ kJ} = -49\,306 \text{ kJ}$

$= -49.3 \text{ MJ}$

Hence the combustion enthalpy of pentane is −49.3 MJ kg^{-1}.

SAQ 6.3 (Objective 6.3)
Convert to MJ kg^{-1} the combustion enthalpies of:
(a) hydrogen
(b) ethylene.

Use the data in Table 6.1. The relative atomic masses of hydrogen and carbon are 1 and 12 respectively.

If you look at Table 6.1, you'll see that I've been careful to specify the state of the system in terms of whether the reactants and products are solid, liquid or gas (denoted by the symbols s, l or g), and the temperatures of the reactants before the reaction and of the products after it. The reasons for this are explored in ▼Specifying states — the standard state▲.

▼Specifying states — the standard state▲

Why do you need to specify states? Consider the simple combustion reaction

$$H_2(g) + \tfrac{1}{2} O_2(g) \longrightarrow H_2O(l)$$

Suppose that initially the hydrogen and oxygen gases are at a temperature of 298 K and that after the reaction, the water is cooled down from the reaction temperature to a temperature of 298 K. These are precisely the experimental conditions referred to in Table 6.1, and so you would expect the combustion enthalpy to be -286 kJ mol^{-1} of H_2 participating in the reaction (the minus sign tells you that it's an exothermic reaction).

Now suppose that the reaction were repeated with the initial reactants again at a temperature of 298 K but that after the reaction, the final product (water) was only cooled down to 350 K instead of 298 K. What effect does this have on the measured enthalpy change? You can estimate the effect from Unit 4. The specific heat capacity of water is $4190 \text{ J kg}^{-1} \text{ K}^{-1}$ and, since 1 mole of water has a mass of 18 g, the molar heat capacity is

$$4190 \times \frac{18}{1000} \text{ J mol}^{-1} \text{ K}^{-1} = 75.4 \text{ J mol}^{-1} \text{ K}^{-1}$$

The energy that would have been released in cooling 1 mole of water from 350 K to 298 K is

    molar heat capacity × temperature change

    $= 75.4 \text{ J mol}^{-1} \text{ K}^{-1} \times (298 - 350) \text{ K}$

    $= -3920 \text{ J mol}^{-1}$

    $\approx -4 \text{ kJ mol}^{-1}$

But this hasn't been released, hence the enthalpy change will now be

$$(-286 + 4) \text{ kJ mol}^{-1} = -282 \text{ kJ mol}^{-1}$$

(Although in this case the difference is less than 1.5%, in other cases the effect can be much more significant.)

EXERCISE 6.2 Calculate the enthalpy change for the combustion of hydrogen gas in oxygen to produce liquid water if the final temperature of the water is 370 K.

Now suppose that the reaction were carried out a third time. This time the reactants and the product are both at 298 K, but now the water remains as a gas (this might happen if the reaction were performed at low pressure). The reaction is now described by a similar equation, but with $H_2O(l)$ replaced by $H_2O(g)$:

$$H_2(g) + \tfrac{1}{2} O_2(g) \longrightarrow H_2O(g)$$

This time the difference in the enthalpy change arises because the latent heat of vaporization of the water is now not released by the system. Again from Unit 4, you should be able to estimate how big this difference will be. The latent heat of vaporization of water is 2.26 MJ kg^{-1} and again, since the relative molecular mass of water is 18, this can be converted to a value per mole:

$$2.26 \times 10^6 \times \frac{18}{1000} \text{ J mol}^{-1}$$

$$= 4.1 \times 10^4 \text{ J mol}^{-1}$$

$$= 41 \text{ kJ mol}^{-1}$$

If therefore the product is gaseous water, then an energy of 41 kJ must be retained by the system for every mole of water in the system. The enthalpy change for this reaction is not -286 kJ mol^{-1} as shown in Table 6.1 but

$$(-286 + 41) \text{ kJ mol}^{-1} = -245 \text{ kJ mol}^{-1}$$

Clearly then, combustion enthalpy, or any reaction enthalpy change, depends upon the temperatures of the reactants and products and upon their physical state (i.e. solid, liquid or gas). These must therefore be stated. In order to simplify the use of data from different sources, it has now been internationally agreed that thermochemical data will be recorded for a specified set of conditions known as the **standard state**.

The standard state is when reactants and products are in their normal states (e.g. solid, liquid, or gas) at a temperature of 298.15 K (approximately 25°C) and a pressure of $101.325 \text{ kN m}^{-2}$ (approximately 1 atmosphere). (Note that this standard state differs from 's.t.p.' which was defined in Unit 4.) Standard state data are denoted by a 'plimsoll' $\ominus$ superscript, similar to the Plimsoll line symbol used to indicate the safe loading levels of ships under different conditions. Thus for the combustion of hydrogen, you would write $\Delta H^\ominus = -286 \text{ kJ mol}^{-1}$.

6.2.2 Hess's law

If the only reactions for which thermochemical data could be obtained were those that occur in the restrictive experimental conditions of a calorimeter, there would be many blanks in our knowledge. Some reactions could not be made to happen at all, and some data would be inaccurate because the reaction might proceed in such a way (e.g. too slowly) as to prevent accurate measurement.

Fortunately it isn't necessary to measure all reaction enthalpy changes. Once some experimental data are available it's possible to calculate enthalpy changes for related reactions. The basis for such calculations is the law of conservation of energy (i.e. the first law of thermodynamics).

The application of the first law of thermodynamics to chemical reactions was first postulated in 1840 by Hess and is often referred to as **Hess's law**. This states that:

> The enthalpy change in a chemical reaction is independent of the route taken to convert reactants to products; it depends only on the initial and final states and not on any intermediate states through which the system passes.

The law is illustrated schematically in Figures 6.6 and 6.7. You can see now why ΔH is used rather than ΔQ — the law can only work with a state function.

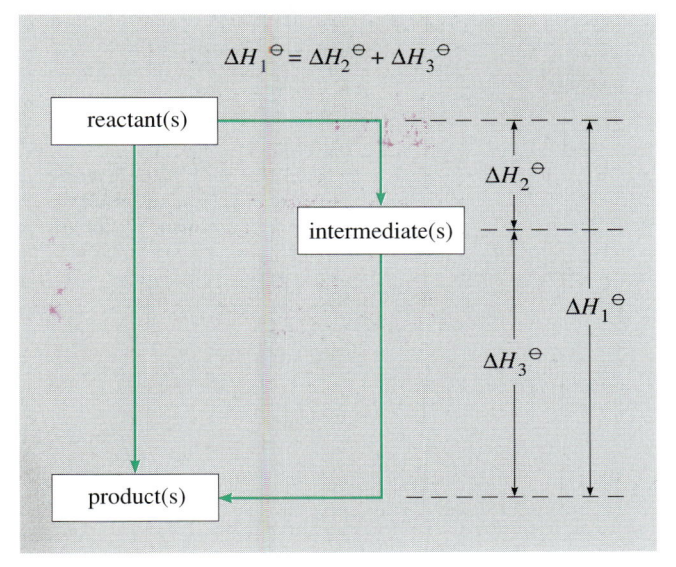

Figure 6.6 Schematic of general application of Hess' law

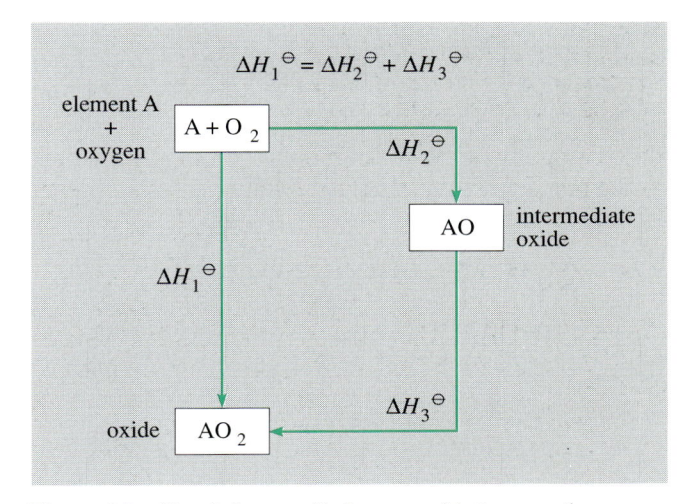

Figure 6.7 Hess's law applied to an oxidation reaction

Hess's law is invaluable as a means of calculating reaction enthalpies which have not been or cannot be measured experimentally. For example, suppose you wished to know the enthalpy change for converting graphite (carbon) to carbon monoxide (CO). It's impractical to measure this experimentally because when carbon reacts with oxygen at constant pressure, the ensuing reaction usually produces carbon dioxide (CO_2):

$$C(s) + O_2(g) \longrightarrow CO_2(g) \qquad \text{(i)}$$
$$\Delta H^\ominus = -394 \text{ kJ (mol C)}^{-1}$$

(See 'Specifying states — the standard state' if you don't know what the '$\ominus$' in $\Delta H^\ominus$ means. Notice also that in this, and in many of the subsequent reactions, the $\Delta H^\ominus$ values are expressed in terms of per mole of a specified reactant — this is especially important when unequal numbers of moles are involved.) It's equally simple to burn carbon monoxide in oxygen so that it is all converted to carbon dioxide:

$$CO(g) + \tfrac{1}{2} O_2(g) \longrightarrow CO_2(g) \qquad \text{(ii)}$$
$$\Delta H^\ominus = -284 \text{ kJ (mol CO)}^{-1}$$

What you want is $\Delta H^\ominus$ for the reaction:

$$C(s) + \tfrac{1}{2} O_2(g) \longrightarrow CO(g) \qquad \text{(iii)}$$

But this reaction can be expressed as the sum of the reactions:

$$C(s) + O_2(g) \longrightarrow CO_2(g) \qquad \text{(i)}$$
$$\Delta H^\ominus = -394 \text{ kJ (mol C)}^{-1}$$

$$CO_2(g) \longrightarrow CO(g) + \tfrac{1}{2} O_2(g) \qquad \text{(iv)}$$
$$\Delta H^\ominus = +284 \text{ kJ (mol CO}_2)^{-1}$$

where reaction (iv) (the decomposition of CO_2) is just the reverse of reaction (ii). When a reaction is reversed the sign of the enthalpy change, $\Delta H^\ominus$ must also be reversed.

If reactions (i) and (iv) are added, you get:

$$C(s) + O_2(g) + CO_2(g) \longrightarrow CO_2(g) + CO(g) + \tfrac{1}{2} O_2(g)$$

or, cancelling out $CO_2(g)$ and $\tfrac{1}{2} O_2(g)$

$$C(s) + \tfrac{1}{2} O_2(g) \longrightarrow CO(g)$$

which is the required reaction (iii), which has a net enthalpy change of

$$\Delta H^\ominus = (-394 + 284) \text{ kJ (mol C)}^{-1}$$
$$= -110 \text{ kJ (mol C)}^{-1}$$

SAQ 6.4 (Objectives 6.3 and 6.4)
When solid sulphur is burned in oxygen it produces sulphur dioxide (SO_2), for which the standard enthalpy of the reaction is -298 kJ (mol S)$^{-1}$. Sulphur dioxide can be reacted with oxygen to produce sulphur trioxide (SO_3), with a standard reaction enthalpy of -99 kJ (mol SO_2)$^{-1}$. Calculate the reaction enthalpy when solid sulphur is converted to sulphur trioxide under standard conditions.

SAQ 6.5 (Objectives 6.3 and 6.4)
Sulphur can exist in two different crystalline or allotropic forms, called monoclinic sulphur and rhombic sulphur. When burned in oxygen both forms produce sulphur dioxide and both reactions are exothermic with reaction enthalpies of -298.3 kJ mol^{-1} for monoclinic and -298.0 kJ mol^{-1} for rhombic sulphur. Calculate the enthalpy change occurring if rhombic sulphur is converted to monoclinic sulphur. Assume that all reactants and products are in their standard states.

6.2.3 Enthalpies of formation

So far you've seen the enthalpy changes that occur when chemical reactions take place. These changes are of direct interest to engineers as well as chemists and they can be directly measured. However, there are millions of chemical compounds, many of which can interact chemically with one another, so there is a vast number of possible reactions. If, therefore, thermochemical data were tabulated as reaction enthalpies, the resulting tables would be unmanageably large.

However, because Hess's law allows you to calculate reaction enthalpy changes from related reactions, it's not necessary to tabulate data for every reaction. A convenient way of summarizing thermochemical data is as **enthalpy of formation**. This is the enthalpy change occurring when a compound is formed from its constituent elements and is usually denoted by the symbol $\Delta H_f^\ominus$. The subscript 'f' denotes 'formation' and the 'plimsoll' sign denotes the standard states of reactants and products (which you've seen is: temperature 298.15 K, pressure 101.325 kN m^{-2}).

> The enthalpies of formation of the elements in their standard states are defined to be zero.

For instance, the standard state of hydrogen is $H_2(g)$ and the standard state of oxygen is $O_2(g)$. When an element can exist in more than one form (e.g. carbon as graphite or diamond) it's usual to specify which form is used.

SCIENCE FOR MATERIALS

Some enthalpies of formation can be measured directly; see, for example, the first entry in Table 6.1 — this represents the formation of water. Many, however, refer to hypothetical reactions which have been calculated from measured reactions using Hess's law. Enthalpies of formation are widely tabulated in reference books and Tables 6.2 and 6.3 give values for selected inorganic and organic compounds respectively.

You can use enthalpies of formation to calculate reaction enthalpy changes. If you recall Figure 6.4, you can express this in terms of enthalpies of formation as shown in Figure 6.8. The net enthalpy of the reactants is the sum of their standard enthalpies of formation, and similarly for the products.

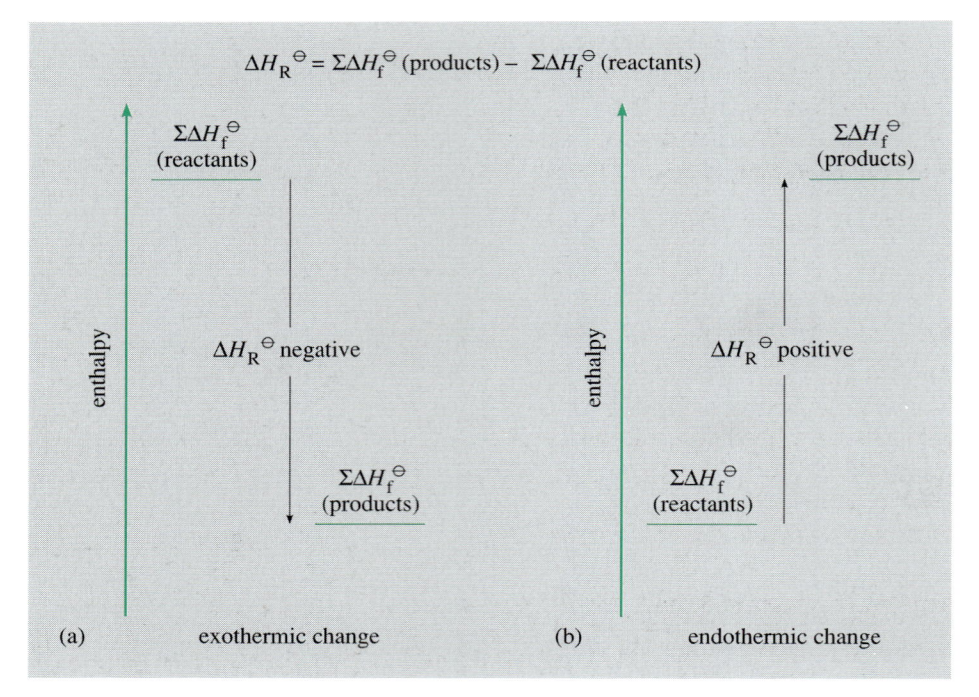

Figure 6.8 Schematic of reaction enthalpy in terms of standard enthalpies of formation of reactants and products

The difference between the net enthalpy of the products and the net enthalpy of the reactants is the standard enthalpy of the reaction $\Delta H_R^\ominus$. In other words, the overall reaction enthalpy per mole of a specified reactant is the sum of the formation enthalpies of the products, minus the sum of the formation enthalpies of the reactants, or, mathematically (see 'Signs of algebra II' in Unit 3):

$$\Delta H_R^\ominus = \sum_{\text{products}} \Delta H_f^\ominus - \sum_{\text{reactants}} \Delta H_f^\ominus \tag{6.3}$$

The only thing you've got to watch out for is that you start with a properly balanced chemical equation. Then the same multipliers as in that equation are used to multiply the individual $\Delta H_f^\ominus$ terms in Equation (6.3).

Suppose, for instance, that you wanted to calculate the enthalpy change when methane (CH_4) gas is burned completely in oxygen. This is what happens when you light a gas cooker if it's supplied with 'natural gas'. The balanced chemical reaction is

$$CH_4(g) + 2O_2(g) \longrightarrow CO_2(g) + 2H_2O(l)$$

From Tables 6.2 and 6.3, the standard formation enthalpies for methane, carbon dioxide and water are:

$CH_4(g)$ $\Delta H_f^\ominus = -74.8 \text{ kJ mol}^{-1}$

$CO_2(g)$ $\Delta H_f^\ominus = -393.5 \text{ kJ mol}^{-1}$

$H_2O(l)$ $\Delta H_f^\ominus = -285.9 \text{ kJ mol}^{-1}$

For the $O_2(g)$, you should remember that the enthalpy of formation of the elements in their standard state is zero, so

$O_2(g)$ $\Delta H_f^\ominus = 0$

Applying Equation (6.3)

$$\Delta H_R^\ominus = \sum_{\text{products}} \Delta H_f^\ominus - \sum_{\text{reactants}} \Delta H_f^\ominus$$

$$(\Delta H_f^\ominus(CO_2) + 2\Delta H_f^\ominus(H_2O)) - (\Delta H_f^\ominus(CH_4) + \Delta H_f^\ominus(O_2))$$

$$= (-393.5 + (2 \times -285.9)) - (-74.8 + 0) \text{ kJ mol}^{-1}$$

$$= -890.5 \text{ kJ mol}^{-1}$$

This is the enthalpy change when one mole of CH_4 is burned, so you can say

$$\Delta H_R^\ominus = -891 \text{ kJ (mol } CH_4)^{-1}$$

Notice that, if you'd wanted the enthalpy change per mole of oxygen, the appropriate equation would have been

$$\tfrac{1}{2}CH_4(g) + O_2(g) \longrightarrow \tfrac{1}{2}CO_2(g) + H_2O(l)$$

which would have provided a value only one-half that calculated above.

Table 6.2 Standard enthalpies of formation for selected inorganic compounds

Compound	Formula	State	$\Delta H_f^\ominus / \text{kJ mol}^{-1}$
aluminium chloride	$AlCl_3$	s	−705.3
aluminium oxide	Al_2O_3	s	−1673.6
barium sulphate	$BaSO_4$	s	−1465.2
carbon monoxide	CO	g	−110.5
carbon dioxide	CO_2	g	−393.5
carbon disulphide	CS_2	l	+87.9
calcium chloride	$CaCl_2$	s	−795.8
calcium oxide	CaO	s	−635.1
calcium carbonate	$CaCO_3$	s	−1206.9
copper (I) oxide, cuprous oxide	Cu_2O	s	−166.7
copper (II) oxide, cupric oxide	CuO	s	−155.8
iron (II) oxide, ferrous oxide	FeO	s	−266.5
iron (III) oxide, ferric oxide	Fe_2O_3	s	−822.2
iron oxide, magnetite	Fe_3O_4	s	−1118.4
hydrogen chloride	HCl	g	−92.3
hydrogen chloride	HCl	aq	−167.4
water	H_2O	l	−285.9
water	H_2O	g	−241.8
hydrogen sulphide	H_2S	g	−20.5
mercury (I) oxide, mercurous oxide	Hg_2O	s	−91.2
mercury (II) oxide, mercuric oxide	HgO	s	−90.8
magnesium oxide	MgO	s	−601.7
magnesium carbonate	$MgCO_3$	s	−1095.8
ammonia	NH_3	g	−46.0
ammonium chloride	NH_4Cl	s	−314.5
nitrogen (I) oxide, nitrous oxide	N_2O	g	+81.5
nitrogen (II) oxide, nitric oxide	NO	g	+90.3
sodium hydroxide	$NaOH$	s	−428.0
sodium carbonate	Na_2CO_3	s	−1130.8
phosphorus pentoxide	P_2O_5	s	−2984.0
sulphur dioxide	SO_2	g	−298.3
sulphur trioxide	SO_3	g	−395.8
sulphuric acid	H_2SO_4	l	−811.3
sulphuric acid	H_2SO_4	aq	−907.5
silicon dioxide	SiO_2	s	−910.9
zinc oxide	ZnO	s	−348.1

Table 6.3 Standard enthalpies of formation for selected organic compounds

Compound	Formula	State	$\Delta H_f^\ominus$ / kJ mol^{-1}
methane	CH_4	g	−74.8
ethane	C_2H_6	g	−84.7
propane	C_3H_8	g	−103.8
butane	C_4H_{10}	g	−126.2
pentane	C_5H_{12}	g	−134.5
cyclohexane	C_6H_{12}	l	−156.2
ethylene (ethene)	C_2H_4	g	+52.3
propylene (propene)	C_3H_6	g	+20.4
acetylene (ethyne)	C_2H_2	g	+226.7
methanol	$CH_3.OH$	l	−238.6
ethanol	$C_2H_5.OH$	l	−277.0
acetaldehyde (ethanal)	$CH_3.CHO$	g	−166.4
dimethyl ketone (2-propanone)	$(CH_3)_2.CO$	l	−248.1
acetic acid (ethanoic acid)	$CH_3.COOH$	l	−484.1

SAQ 6.6 (Objectives 6.3 and 6.4)
Acetylene gas (C_2H_2) burns completely in oxygen to produce carbon dioxide gas and water as liquid. If reactants and products are in their standard states, calculate the reaction enthalpy change for this combustion reaction. Use the data in Tables 6.2 and 6.3.

SAQ 6.7 (Objectives 6.3 and 6.4)
In oil refining, a cracker operates by heating saturated hydrocarbons in the absence of air, when they decompose into unsaturated hydrocarbons of lower molecular mass. One such reaction is the decomposition of pentane into ethylene according to the reaction:

$$C_5H_{12}(g) \longrightarrow 2C_2H_4(g) + CH_4(g)$$

Use the data of Table 6.3 to calculate the enthalpy change during this reaction and hence state whether it is exothermic or endothermic.

6.2.4 Entropy changes in phase changes and reactions

You saw in Unit 4, Section 4.6.3 that, at phase changes, ΔH and $T\Delta S$ are just in balance and $\Delta G = 0$. For instance, on changing from solid to liquid at T_m, the enthalpy required by the melting is just sufficient to compensate for the increase in entropy as the system gets more disordered. Thus, in terms of the free energy equation (Equation 6.1), you can write for melting (subscript 'm') under standard conditions:

$$\Delta G_m^\ominus = \Delta H_m^\ominus - T_m \Delta S_m^\ominus = 0$$

This can be rearranged to

$$T_m = \frac{\Delta H_m^\ominus}{\Delta S_m^\ominus} \tag{6.4}$$

$\Delta H_m^\ominus$ is just the molar latent heat of fusion of the material, which, together with T_m can readily be determined, as you saw in Unit 4. Values of latent heat and melting temperature for a range of metals are shown in Table 6.4, together with the calculated entropy changes that occur on melting.

Table 6.4 Melting characteristics of some metals

Material	$\dfrac{T_m}{K}$	$\dfrac{\Delta H_m^\ominus}{\text{kJ mol}^{-1}}$	$\dfrac{\Delta S_m^\ominus}{\text{J mol}^{-1}\text{K}^{-1}}$
aluminium	933	10.79	11.58
copper	1356	13.10	9.65
iron	1811	15.54	8.62
tungsten	3650	35.36	9.69
mercury	234	2.31	9.87
titanium	1998	18.90	9.46
magnesium	923	9.07	9.83
tin	505	7.22	14.30

From these calculated values for the entropy of fusion you can see that even though the melting temperature of the different metals varies widely (compare mercury with tungsten), the entropies of fusion all lie fairly close to one another. As entropy is related to the amount of disorder, the increase in disorder on melting is very similar for all metals.

The other point to notice from Equation (6.4) is that the entropy change $\Delta S_m^\ominus$ and the enthalpy of fusion $\Delta H_m^\ominus$ are directly related. Since T_m will always be positive, it follows that both $\Delta H_m^\ominus$ and $\Delta S_m^\ominus$ will always be of the same sign. Melting only occurs when thermal energy is put into a system; that is, it is an endothermic process and so $\Delta H_m^\ominus$ will be positive. $\Delta S_m^\ominus$ will therefore also be positive. In other words, there is always an increase in disorder on melting.

Solidification is the reverse of melting and occurs only when thermal energy is removed from the system. The equation describing solidification will be identical to Equation (6.4) but now $\Delta H_m^\ominus$ will be negative and the entropy change $\Delta S_m^\ominus$ will also be negative, indicating an increase in the order of the material comprising the system.

This treatment of melting can be extended to describe the changes at boiling and sublimation. (Remember that in boiling (or vaporization) a liquid turns to a gas and in sublimation a solid changes directly to a gas.) Both of these types of change are also described by Equation (6.4) if the subscript 'm' is changed to 'b' to denote boiling, or 's' to denote sublimation.

SAQ 6.8 (Objectives 6.3, 6.4 and 6.5)
Table 6.5 gives the boiling (vaporization) temperatures T_b and the enthalpies of vaporization $\Delta H_v^\ominus$ of a number of metals.

(a) Calculate the entropies of vaporization of the metals and enter your results in the table.

(b) How do these entropies compare with the entropies of fusion for the same metal (Table 6.4)?

(c) Account for the difference between the entropy of fusion and the entropy of vaporization of the metals (remember Unit 4).

Table 6.5 Vaporization characteristics of some metals

Material	T_b K	$\Delta H_v^\ominus$ kJ mol^{-1}	$\Delta S_v^\ominus$ J mol^{-1} K^{-1}
aluminium	2740	285.4	
copper	2868	305.8	
iron	3008	355.3	
tungsten	6200	775.3	
mercury	630	59.2	
titanium	3530	424.2	
magnesium	1380	132.3	
tin	2540	291.5	

6.2.5 Free energy changes in reactions

Just as standard entropies and enthalpy changes can be calculated from the standard values, it is possible to calculate standard free energy changes for reactions using tabulated standard entropies and enthalpies.

Why is this important?

Since it is the free energy change that determines the feasibility of a reaction occurring, it is possible to predict whether or not *any* compound will be stable under given temperatures and pressures. This predictive value of thermodynamics is useful, for instance, in deciding whether a particular reaction route is a feasible way to synthesize a new compound. The condition is that ΔG for any reaction by which a new compound or material is formed must be negative. You'll appreciate from the previous discussion on the conservation of energy that Hess's law must be applicable to entropy and free energy changes just as it is to enthalpy changes, so that a great deal of useful information (sometimes for hypothetical reactions) can be gathered from a limited amount of experimental data.

There's a second reason why measurement of free energy changes is important. Many processes for making materials were originally developed by trial and error. Applying thermodynamics theory can show how energy-efficient these processes are, and can point the way to alternative, more efficient processes. For example, in steelmaking, blast furnaces don't use energy very efficiently. Now, in some countries (often where there's a plentiful supply of cheap hydroelectric power) 'direct reduction' processes are used instead.

Free energy information for reactions is often presented in the form of free energy diagrams, based on the free energy equation. If you compare the equation

$$\Delta G = \Delta H - T\Delta S$$

with the standard form of a straight line equation

$$y = mx + c$$

you should see that, if ΔG varies linearly with T, a plot of ΔG against T will give you a straight line of slope $(-\Delta S)$ and intercept ΔH (Figure 6.9). Note that ΔS is negative, so the slope is positive. Many reactions behave in this way, implying that ΔS is independent of temperature. Note that since most reactions occur at 1 atmosphere, you can use the standard notation immediately. What does this diagram imply for a chemical reaction?

Since $\Delta G^\ominus$ must be negative for a reaction to occur, the region *below* the horizontal axis is where the reaction occurs. Above the temperature axis, $\Delta G^\ominus$ is positive and the reaction cannot occur at one atmosphere pressure. The point at which the free energy line crosses the axis is known as a critical temperature (T_c) and marks the point at which the reaction stops. It also marks the point where the reverse reaction becomes thermodynamically feasible.

The form of free energy diagrams for reactions at constant pressure is standardized in several ways:

- $\Delta G^\ominus$ is the vertical axis and T the horizontal axis.
- Reactions are expressed in terms of a standard amount of one of the reactants. (e.g. 1 mole of oxygen when considering oxide formation).
- Reactions are always written as proceeding in the direction: element $\longrightarrow$ compound.

Their single most important use is for metal–metal oxide systems where they are known as **Ellingham diagrams** (after the individual who first used them, around 1948). A simple example is shown in Figure 6.10. This reaction is the classic one used by Joseph Priestley when he first discovered oxygen in 1774 (see ▼Priestley and oxygen▲ on p.88). There is more on Ellingham diagrams and their application in the next Block, but you can see

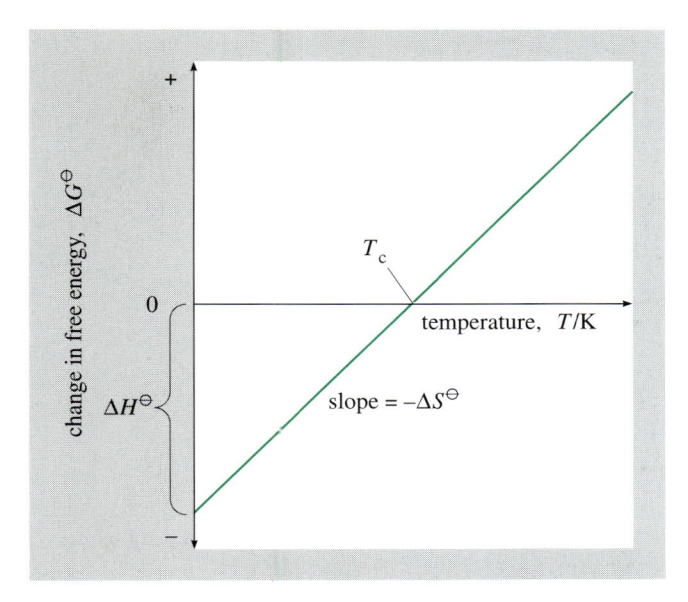

Figure 6.9 Schematic free energy diagram

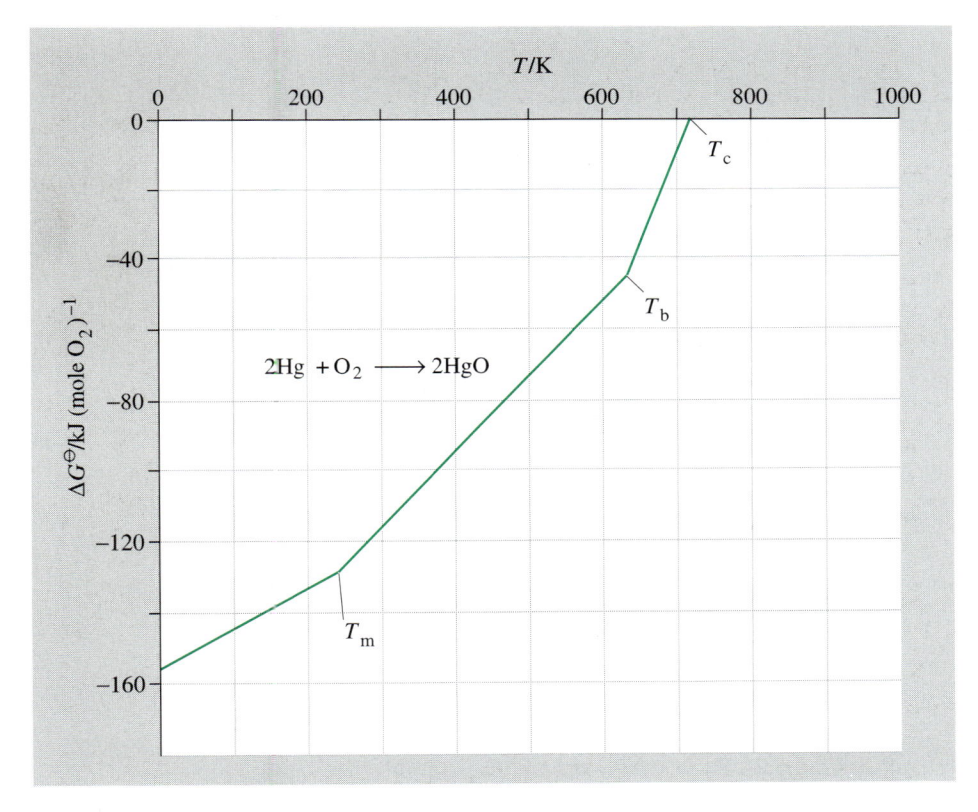

Figure 6.10 Ellingham diagram for mercury–mercuric oxide

one feature in Figure 6.10 that you'll find in many diagrams. The line consists of three segments of different slopes. You'll see later that there is a phase change associated with each change in slope — in this case melting (at T_m) followed by boiling (at T_b).

SCIENCE FOR MATERIALS

▼Priestley and oxygen▲

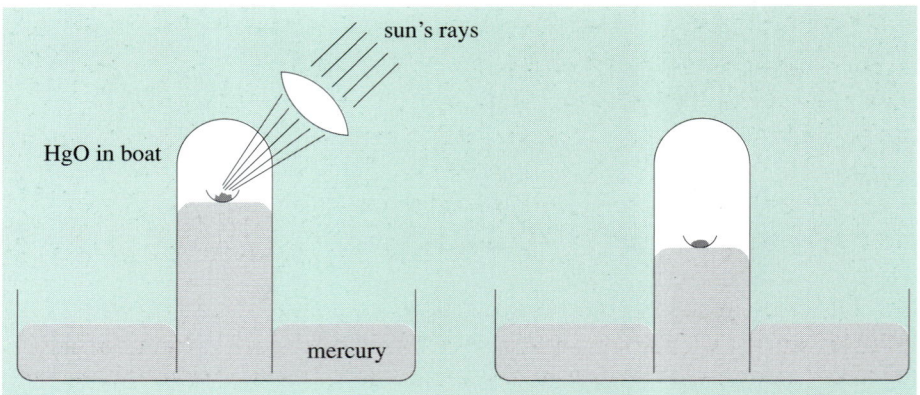

Figure 6.12 Priestley's experiment for making oxygen gas. The Sun's rays were focussed on mercuric oxide in a boat floating on mercury in a bell jar (left); the resulting decomposition produced metallic mercury and oxygen gas (right)

Figure 6.11 Priestley, a contemporary view

The experiment that led Joseph Priestley (Figure 6.11) to the discovery of oxygen was very simple : a convex lens, red mercuric oxide (HgO) and a flask inverted over a trough of mercury (Figure 6.12). In his own words:

'… having procured a lens of twelve inches diameter, and twenty inches focal distance, I proceeded with great alacrity to examine, by the help of it, what kind of air a great variety of substances, natural and factitious [artificially prepared] would yield … With this apparatus on the 1st August, 1774, I endeavoured to extract air from *mercurius calcinatus per se* [red mercuric oxide]; and I presently found that, by means of this lens, air was expelled from it very readily. Having got about three or four times as much as the bulk of my materials, I admitted water to it, and found that it was not imbibed by it. But what surprised me more than I can well express, was, that a candle burned in this air with a remarkably vigorous flame … I was utterly at a loss how to account for it.'

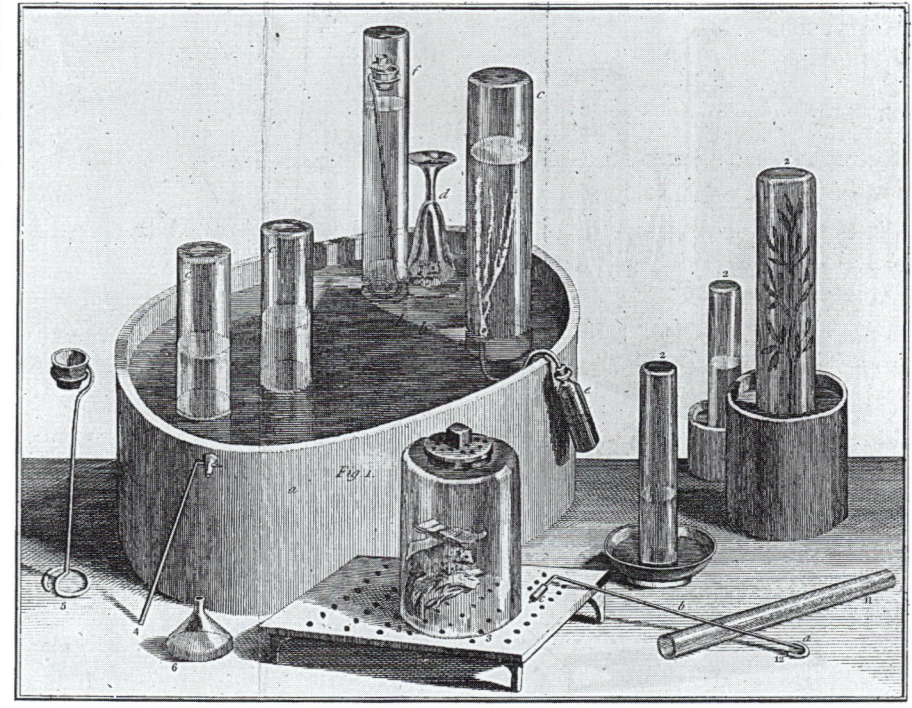

Figure 6.13 Priestley's apparatus for various bell-jar experiments using oxygen (original experiment, upper left centre; photosynthesis, extreme right; mice in oxygen, lower centre)

He breathed the gas himself and found his '… breast felt peculiarly light and easy for some time afterwards'. He recommended its use in medicine, which is, of course, now taken for granted since oxygen is vital for respiration of the blood (e.g. the mice in Figure 6.13). He also suggested that the gas could be used to achieve high temperatures when fed directly to a flame, another commonplace today with oxy-acetylene and oxy-hydrogen burners.

SAQ 6.9 (Objective 6.6)
Based on Figure 6.10:
(a) What is the temperature above which mercuric oxide decomposes to mercury and oxygen gas?
(b) What is the standard reaction enthalpy per mole of mercury for the formation of its oxide?
(c) What is the standard entropy change per mole of oxygen for the oxidation of liquid mercury?

Ellingham diagrams are especially useful for showing the critical temperatures at which you can start converting ores such as metal oxides into metals. The region where ΔG is positive in the diagram is where ΔG for the reverse reduction reaction (i.e. oxide $\longrightarrow$ metal) is negative (SAQ 6.9(a)). Table 6.6 lists the critical temperatures for some metal–metal oxide systems. This shows the high temperatures required by many such direct routes to extract metals, and explains why alternative reaction routes are often adopted, for instance the use of carbon to provide a reducing agent in the blast furnace.

Table 6.6 Critical temperatures for reduction of some metal oxides

Reaction	T_c / K
$2HgO(s) \longrightarrow 2Hg(s) + O_2(g)$	710
$2CuO(s) \longrightarrow 2Cu(s) + O_2(g)$	1710
$2Ag_2O(s) \longrightarrow 4Ag(s) + O_2(g)$	1800
$2Cu_2O(s) \longrightarrow 4Cu(s) + O_2(g)$	2750
$2MgO(s) \longrightarrow 2Mg(s) + O_2(g)$	3600
$WO_2(s) \longrightarrow W(s) + O_2(g)$	3850
$2FeO(s) \longrightarrow 2Fe(s) + O_2(g)$	5400

Summary

- Combustion enthalpy is the enthalpy change when an element or compound reacts completely with oxygen.
- Hess's law says that any reaction enthalpy change depends only on the final and initial states; it is independent of the reaction route or any intermediate states.
- The standard enthalpy of formation $\Delta H_f^\ominus$ of a compound is the enthalpy change when the compound is formed from its constituent elements under standard conditions.
- By definition, the enthalpies of formation of the elements in their standard states are zero.
- The standard enthalpy change of a reaction, $\Delta H_R^\ominus$, is given by

$$\Delta H_R^\ominus = \sum_{\text{products}} \Delta H_f^\ominus - \sum_{\text{reactants}} \Delta H_f^\ominus$$

- The entropy change associated with a phase change can be determined from the free energy equation, with $\Delta G = 0$.
- Free energy (or Ellingham) diagrams are plots of ΔG versus T for a reaction, and show the temperature ranges for which the reaction or its reverse are feasible.

SCIENCE FOR MATERIALS

6.3 Equilibrium reactions

6.3.1 The hydrogen iodide reaction

Oxidation reactions like that of mercury in Figure 6.10 are reactions where $\Delta G^\ominus$ is usually large and negative at 298 K, so once they're started, they go to completion. In other words, all the reactants are converted to products. As you've already seen, this is not always the case. Many reactions proceed up to a certain point and then don't change any further, no matter how long you wait. The solubility of salt in water provided a simple example of such an equilibrium reaction in Section 6.1.1.

Now I want to take a more detailed look at an even simpler reaction — the formation of hydrogen iodide (HI) gas. Although this reaction isn't directly relevant to materials engineering, it is useful as a simple model to begin the exploration of equilibrium reactions in general.

Hydrogen iodide gas is formed by reaction between gaseous hydrogen (H_2) and iodine (I_2). It is easy to study because one component (the iodine) is coloured, so its presence can be used to follow the reaction. All components are gaseous, so there are no problems associated with surface reactions or other characteristics of condensed matter (solids and liquids).

When hydrogen and iodine gases are heated in a glass vessel to a temperature of 700 K, the gas mixture initially exhibits the deep purple colour characteristic of iodine gas. If the vessel is maintained at 700 K, the colour fades but settles down to a steady shade after about an hour. An analysis of the gas mixture at this time would show that it contains all three of hydrogen, iodine and hydrogen iodide. Alternatively, if the vessel is initially filled with colourless hydrogen iodide gas and heated to 700 K, it is now found that it gradually acquires a purple tint which again settles down to a steady shade after about an hour. Analysis of the contents would show that the hydrogen iodide had partially decomposed into hydrogen and iodine. Thus the two reactions must have been

$$H_2(g) + I_2(g) \xrightarrow{700\,K} 2HI(g)$$

and

$$2HI(g) \xrightarrow{700\,K} H_2(g) + I_2(g)$$

So both these opposing reactions can occur at 700 K; the reaction of hydrogen with iodine to produce hydrogen iodide, and the decomposition of hydrogen iodide to produce hydrogen and iodine. Since both can occur in the reaction vessel, the equilibrium state is explained as that which occurs when the rate of formation of hydrogen iodide is equal to its rate of decomposition. When this condition is reached the reactions are in dynamic equilibrium. You've seen that such equilibrium reactions are written as

$$H_2(g) + I_2(g) \rightleftharpoons 2HI(g)$$

UNIT 6 CHEMICAL REACTIONS

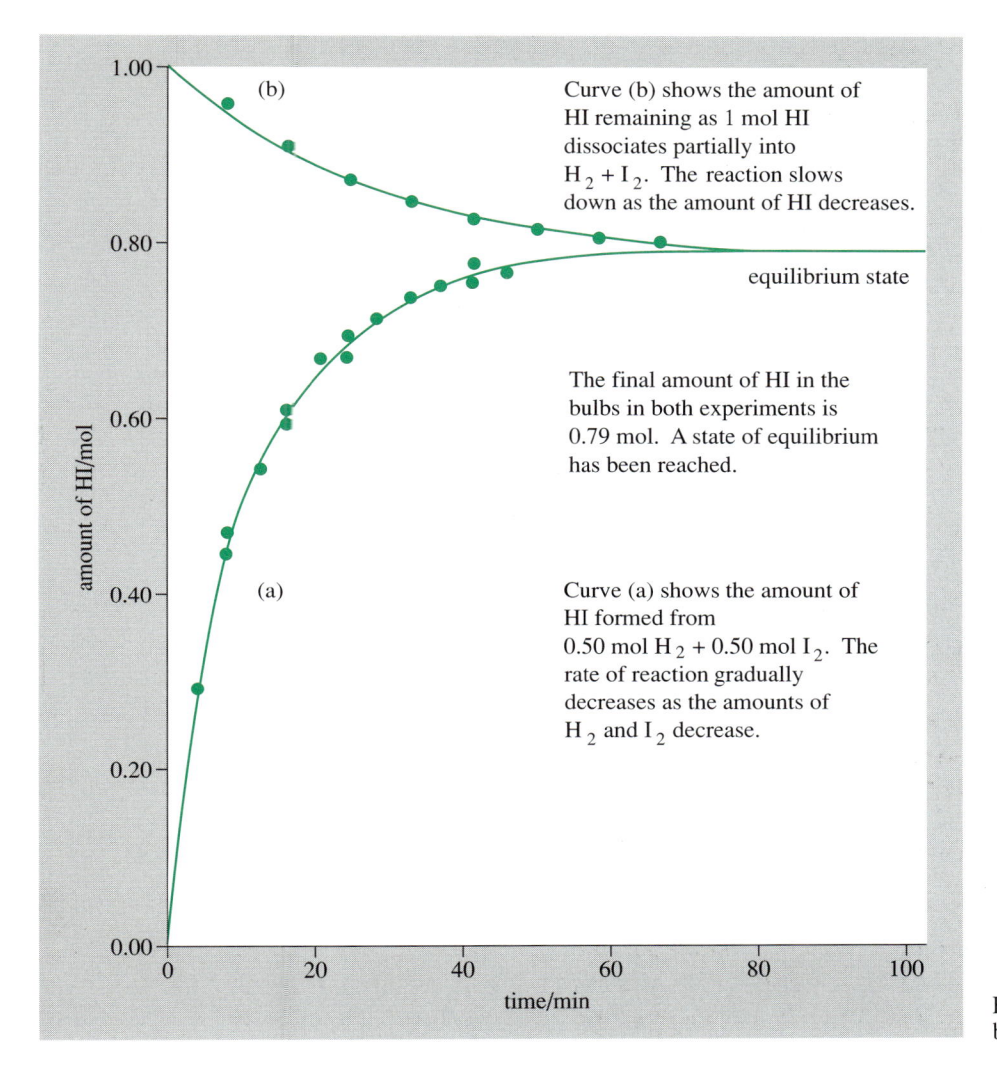

Figure 6.14 (graph): amount of HI/mol vs time/min

- Curve (b) shows the amount of HI remaining as 1 mol HI dissociates partially into $H_2 + I_2$. The reaction slows down as the amount of HI decreases.
- equilibrium state
- The final amount of HI in the bulbs in both experiments is 0.79 mol. A state of equilibrium has been reached.
- Curve (a) shows the amount of HI formed from 0.50 mol H_2 + 0.50 mol I_2. The rate of reaction gradually decreases as the amounts of H_2 and I_2 decrease.

Figure 6.14 The approach to equilibrium between HI, H_2 and I_2 at 700 K

If the concentration of, say, the hydrogen iodide was measured at intervals following the start of each of the two reactions, the results would look like those in Figure 6.14. In Figure 6.14, the lower curve shows how the amount of HI increases in the reaction vessel starting with 0.5 mol each of H_2 and I_2. The upper curve shows the decrease in the amount of HI starting from 1 mol of HI. In both cases the equilibrium amount of HI is the same; namely 0.79 mol.

6.3.2 The equilibrium constant

So how can such an equilibrium be described in a quantitative way? Table 6.7 gives some experimental data obtained for equilibrium mixtures formed when the hydrogen iodide reaction at 700 K is carried out with five different concentrations of starting materials. The convention for expressing concentration is to enclose the chemical formula in square brackets. Thus

[HI] means the concentration of hydrogen iodide, and in Table 6.7 this is measured in units of moles per litre, or mol l^{-1}. The final column in Table 6.7 shows the value of the quantity $[HI]^2/([H_2][I_2])$.

Table 6.7 Equilibrium concentrations of hydrogen, iodine and hydrogen iodide

Experiment number	Experimental results: concentrations at equilibrium			
	$\dfrac{[H_2]}{\text{mol l}^{-1}}$	$\dfrac{[I_2]}{\text{mol l}^{-1}}$	$\dfrac{[HI]}{\text{mol l}^{-1}}$	$\dfrac{[HI]^2}{[H_2][I_2]}$
1	0.48×10^{-3}	0.48×10^{-3}	3.54×10^{-3}	54.4
2	1.14×10^{-3}	1.14×10^{-3}	8.41×10^{-3}	54.4
3	1.83×10^{-3}	3.13×10^{-3}	17.66×10^{-3}	54.4
4	2.91×10^{-3}	1.71×10^{-3}	16.45×10^{-3}	54.4
5	4.56×10^{-3}	0.74×10^{-3}	13.54×10^{-3}	54.3

You can see that $[HI]^2/[H_2][I_2])$ turns out to be a constant and this can be defined by the equation

$$K = \frac{[HI]^2}{[H_2][I_2]} \tag{6.5}$$

K is the **equilibrium constant** for the reaction. Under defined conditions of temperature and pressure, its value is independent of the concentrations of reactants initially present in the reaction vessel.

6.3.3 General definition of equilibrium constant

So for the hydrogen iodide reaction, K is simply defined. But what about the following reaction?

$$2S(g) + 3O_2(g) \rightleftharpoons 2SO_3(g)$$

Its equilibrium constant is:

$$K = \frac{[SO_3]^2}{[S]^2[O_2]^3}$$

When the equilibrium constant is obtained from the concentrations of reactants and products as in Equation (6.5) and Table 6.7, it is often referred to as the **concentration equilibrium constant** and is given the symbol K_c. Thus Equation (6.5) becomes:

$$K_c = \frac{[HI]^2}{[H_2][I_2]} \tag{6.6}$$

and similarly for the sulphur–oxygen reaction. The reason for this is that the amounts of the participants can be expressed in different ways. For example, in a gaseous reaction, amounts could be expressed in terms of partial pressure of each gas (i.e. the pressure it exerts independently of other gases in the system), and a pressure equilibrium constant K_p could be defined in a very similar way to K_c. Because pressure is measured in different units, K_p can be numerically different from K_c.

You may have spotted the way to determine the powers of the concentration terms; they are simply the coefficients in the balanced chemical reaction. Thus, in the general case of a reaction of the form:

$$a\text{A} + b\text{B} + c\text{C} \rightleftharpoons d\text{D} + e\text{E} + f\text{F},$$

the concentration equilibrium constant is given by:

$$K_c = \frac{[\text{D}]^d[\text{E}]^e[\text{F}]^f}{[\text{A}]^a[\text{B}]^b[\text{C}]^c} \qquad (6.7)$$

This equation is known as the **law of mass action**. As you can see, it's very important to ensure that you start with a balanced chemical equation. The exponents (a to f) in Equation (6.7) are the number of moles of each species in the balanced equation. It should also be apparent that the units of K_c depend on the values of the coefficients (see ▼**Units of equilibrium constant**▲). The higher the value of K_c (or any other equilibrium constant), the further the reaction goes to completion, i.e. the higher the **yield** of the reaction.

6.3.4 Calculating reaction yields

If a reaction of the form

$$\text{A} + \text{B} \longrightarrow \text{C} + \text{D}$$

goes to completion, the yield of products (C, D) can be simply calculated if the number of moles of reactants (A, B) are given. Thus, if one mole each of A and B are reacted together, one mole each of C and D will be formed. But in equilibrium reactions, i.e.

$$\text{A} + \text{B} \rightleftharpoons \text{C} + \text{D}$$

this is no longer valid. However, you can calculate the yield at the reaction equilibrium from the value of the equilibrium constant.

To see this, let's take our old friend

$$\text{H}_2(\text{g}) + \text{I}_2(\text{g}) \underset{}{\overset{700\,\text{K}}{\rightleftharpoons}} 2\text{HI}(\text{g})$$

How do you calculate the yield of hydrogen iodide in this reaction given that $K_c = 54$ at 700 K?

Suppose that 1 mole each of H_2 and I_2 are put into a reaction vessel of volume 1 litre and heated to 700 K. The initial concentration of hydrogen is 1 mol l^{-1} and the initial concentration of iodine is also 1 mol l^{-1} because the volume of the vessel is 1 litre. Let's say that x moles of hydrogen are converted in moving from this initial state to the final state. From the balanced chemical equation for this reaction, if x moles of hydrogen react, then $2x$ moles of HI are formed. So in the equilibrium mixture, the concentrations of hydrogen and iodine will both be $(1 - x)$ mol l^{-1} and that of HI will be $2x$ mol l^{-1}.

▼**Units of equilibrium constant**▲

Concentration equilibrium constant is a product of several terms. The combination of terms appearing in K_c varies from one reaction to another. As a result, the units of equilibrium constant will also vary.

For the hydrogen iodide reaction, where the equilibrium constant K_c is defined by

$$K_c = \frac{[\text{HI}]^2}{[\text{H}_2][\text{I}_2]}$$

[HI], [H_2] and [I_2] are all measured in moles per litre (mol l^{-1}). The units of K_c are therefore given by:

$$\frac{(\text{mol l}^{-1})^2}{(\text{mol l}^{-1})(\text{mol l}^{-1})} = \frac{(\text{mol l}^{-1})^2}{(\text{mol l}^{-1})^2}$$

So the equilibrium constant in this case is dimensionless and is thus just a number.

SAQ 6.10 (Objectives 6.1 and 6.7)
For each of the following, write down an expression for the concentration equilibrium constant, and determine its units.

(a) $C_2H_6(\text{g}) \rightleftharpoons C_2H_4(\text{g}) + H_2(\text{g})$
(b) $2\text{S}(\text{g}) + 3\text{O}_2(\text{g}) \rightleftharpoons 2\text{SO}_3(\text{g})$.

SCIENCE FOR MATERIALS

Now recall that the equilibrium constant K_c is:

$$K_c = \frac{[HI]^2}{[H_2][I_2]}$$

You've seen that, at equilibrium

$[HI] = 2x \text{ mol } l^{-1}$
$[H_2] = [I_2] = (1 - x) \text{ mol } l^{-1}$

Hence

$$K_c = \frac{(2x)^2}{(1 - x)(1 - x)}$$

$$= \frac{(2x)^2}{(1 - x)^2}$$

But $K_c = 54$, therefore

$$\frac{(2x)^2}{(1 - x)^2} = 54$$

Taking the square root of each side gives:

$$\frac{2x}{1 - x} = 7.35$$

or

$$2x = 7.35(1 - x) = 7.35 - 7.35x$$

$$x(2 + 7.35) = 7.35$$

$$x = \frac{7.35}{9.35}$$

$$= 0.79$$

So at equilibrium, the concentrations of hydrogen and iodine will each be

$(1 - 0.79) \text{ mol } l^{-1} = 0.21 \text{ mol } l^{-1}$

and the concentration of HI will be

$(2 \times 0.79) \text{ mol } l^{-1} = 1.58 \text{ mol } l^{-1}$

Complete reaction would have had 1 mole of H_2 reacting with 1 mole of I_2 to give 2 moles of HI. However, the maximum amount of HI produced is only 1.58 mol, or a yield of only about 79% of that expected for complete reaction. This agrees well with the data in Figure 6.14.

SAQ 6.11 (Objective 6.7)
When the reaction $H_2(g) + I_2(g) \rightleftharpoons 2 HI(g)$ is carried out at a temperature of 300 K, the equilibrium constant K_c is 0.040. If 1 mol of H_2 and 1 mol of I_2 are mixed in a reaction vessel of volume 1 litre, estimate the equilibrium yield of the reaction.

Le Chatelier's principle

There are two important points to note in equilibrium calculations. First, for a given set of reaction conditions (i.e. temperature and pressure), the equilibrium constant is not influenced by the initial concentration of the reactants. Rather it depends on the concentrations of the components present in the equilibrium mixture.

The second point is that the initial concentration of reactants does influence the final concentration of products. In the hydrogen iodide reaction, when the initial concentrations of hydrogen and iodine are both $1\,mol\,l^{-1}$, the equilibrium concentration of HI is $1.58\,mol\,l^{-1}$. If the initial concentration of hydrogen is increased to $2\,mol\,l^{-1}$, the [HI] at equilibrium increases to $1.88\,mol\,l^{-1}$. Doubling the initial concentration again to $4\,mol\,l^{-1}$, increases [HI] at equilibrium still further to $1.96\,mol\,l^{-1}$. So if more of one reactant is added, it will shift the equilibrium position so that more of the reactant is converted to product.

This effect was discovered by the French chemist Le Chatelier (1850–1936), who expressed it as a general statement:

> If a system in equilibrium is subjected to some disturbance, the system tends to adjust itself so as to oppose the disturbance and restore equilibrium.

This is known as **Le Chatelier's principle**. In terms of the hydrogen iodide reaction the application of this principle says that if the system were originally set up with $1\,mol\,l^{-1}$ each of hydrogen and iodine, the reaction would proceed until equilibrium was achieved between reactants (H_2 and I_2) and product (HI). If the system is disturbed by the addition of more hydrogen to raise the hydrogen concentration to $2\,mol\,l^{-1}$, the equilibrium concentrations should shift in such a way as to reduce the hydrogen concentration. As you saw above, that's just what happens.

The principle also applies if you change the temperature or pressure of the system. For instance, if you increase the pressure, the position of equilibrium will be shifted towards the side of the reaction that occupies lower volume.

6.3.5 Solutions and solubility

A large number of reactions are carried out in solution. One important group of these comprises reactions between ionic compounds in aqueous solution. You will recall from Unit 5 that ionic compounds are not composed of molecules. Rather, they're collections of ions held together by electrostatic forces. When dissolved in water they readily dissociate. You've seen that, when solid sodium chloride (NaCl) is dissolved in water, the reaction is written as

$$NaCl(s) \xrightleftharpoons{H_2O} Na^+(aq) + Cl^-(aq)$$

and the solution behaves as a liquid in which Na^+ and Cl^- ions are moving around at random. In essence, ions in aqueous solutions behave rather like a gas, except that ions in solution tend to move much more slowly than do the gas molecules that you saw in Unit 4.

Solubility product

When considering solutions it's important to know how much of a substance (the 'solute') will dissolve in a solvent, i.e. what its solubility is. When a saturated solution is in contact with solid solute, there is equilibrium between the two. For the reaction

$$NaCl(s) \underset{}{\overset{H_2O}{\rightleftharpoons}} Na^+(aq) + Cl^-(aq)$$

the equilibrium constant at saturation is

$$K_c = \frac{[Na^+(aq)][Cl^-(aq)]}{[NaCl(s)]}$$

But how do you evaluate [NaCl(s)]? Suppose that a mass m of solid sodium chloride is in equilibrium with the solution. The volume occupied by this solid sodium chloride is (m/ρ), where ρ is its density. If its relative molecular mass is M_m, then a mass m (in grams) is equivalent to (m/M_m) moles. Now, the concentration of solid sodium chloride is the molar mass divided by the volume:

$$[NaCl(s)] = \frac{m/M_m}{m\rho} = \frac{\rho}{M_m} \tag{6.8}$$

But both ρ and M_m are constants, so [NaCl(s)] must also be a constant. In other words, the concentration of the solid does not depend upon how much undissolved solid is present in the solution (as long as *some* solid is present to establish equilibrium). So we can say that [NaCl(s)] is equal to a constant C, and K_c becomes:

$$K_c = \frac{[Na^+(aq)][Cl^-(aq)]}{C}$$

$$CK_c = [Na^+(aq)][Cl^-(aq)]$$

The product CK_c is itself a constant known as the **solubility product** whose symbol is K_{sp}. So

$$K_{sp} = [Na^+(aq)][Cl^-(aq)]$$

Because the terms on the right-hand side of this equation are ions in solution, it is usual to drop the (aq) symbol to give

$$K_{sp} = [Na^+][Cl^-]$$

You should remember that the right-hand side of this equation is derived from the equilibrium law. Thus the indices are determined from the coefficients in the chemical equation. For example, calcium chloride ($CaCl_2$) dissolves as

$$CaCl_2(s) \underset{}{\overset{H_2O}{\rightleftharpoons}} Ca^{2+}(aq) + 2Cl^-(aq)$$

The solubility product will therefore be:

$$K_{sp} = [Ca^{2+}][Cl^-]^2$$

> **SAQ 6.12** (Objectives 6.1 and 6.8)
> For each of the following compounds: (i) write a balanced chemical equation showing the ions produced when they dissolve in water; (ii) write an expression for solubility product in terms of ion concentrations; (iii) determine the units of solubility product if all concentrations are measured in mol l^{-1}.
> (a) Potassium chloride, KCl
> (b) Barium sulphate, BaSO$_4$
> (c) Calcium hydroxide, Ca(OH)$_2$

Calculating solubility product

Solubility product can readily be calculated from the solubility of a compound in water; that is, the mass needed to produce a saturated solution. For instance, at room temperature, no more than 2.4×10^{-3} g of barium sulphate (BaSO$_4$) will dissolve in a litre of water. This is actually a very low solubility. In solution, the BaSO$_4$ dissociates as

$$BaSO_4(s) \xrightleftharpoons{H_2O} Ba^{2+}(aq) + SO_4^{2-}(aq)$$

The situation which occurs is shown in Figure 6.15. As more BaSO$_4$ is added, the so-called ion product, which is the product of $[Ba^{2+}]$ and $[SO_4^{2-}]$, rises from zero to its equilibrium value of K_{sp}.

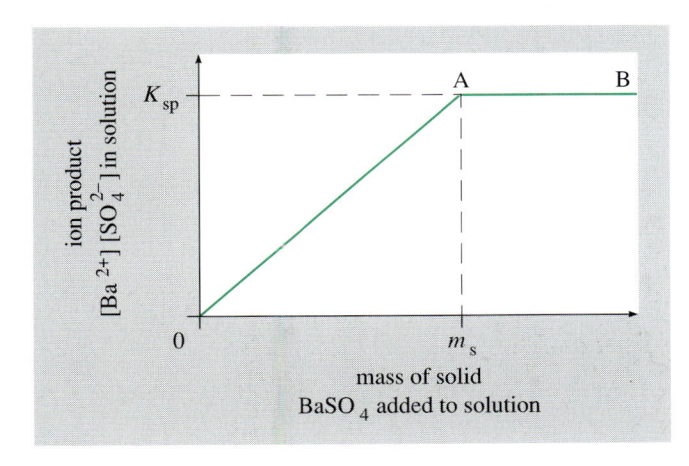

Figure 6.15 Ion product $[Ba^{2+}][SO_4^{2-}]$ in aqueous solution as a function of mass of solid BaSO$_4$ added. m_s represents the solubility of BaSO$_4$

So how do you determine K_{sp} given an equilibrium concentration of 2.4×10^{-3} g l^{-1}? The relative atomic masses of barium, sulphur and oxygen are 137.34, 32.06 and 16.00 respectively, so the molecular mass of barium sulphate is:

$$137.34 + 32.06 + (4 \times 16.00) = 233.4$$

Hence $2.4 \times 10^{-3}\,\text{g}\,\text{l}^{-1}$ of barium sulphate is equivalent to:

$$\frac{2.4 \times 10^{-3}}{233.4}\,\text{mol}\,\text{l}^{-1} = 1.03 \times 10^{-5}\,\text{mol}\,\text{l}^{-1}$$

and in solution this will produce a concentration of $(1.03 \times 10^{-5})\,\text{mol}\,\text{l}^{-1}$ of Ba^{2+} and $(1.03 \times 10^{-5})\,\text{mol}\,\text{l}^{-1}$ of SO_4^{2-}. The solubility product K_{sp} is

$$K_{sp} = [Ba^{2+}][SO_4^{2-}]$$
$$= (1.03 \times 10^{-5})(1.03 \times 10^{-5})\,(\text{mol}\,\text{l}^{-1})^2$$
$$= 1.06 \times 10^{-10}\,\text{mol}^2\,\text{l}^{-2}$$

SAQ 6.13 (Objectives 6.1 and 6.8)
The solubility product of calcium sulphate ($CaSO_4$) is $2.0 \times 10^{-5}\,\text{mol}^2\,\text{l}^{-2}$. Calculate the concentration (in $\text{mol}\,\text{l}^{-1}$) of calcium sulphate in a saturated solution.

6.3.6 Acids and bases

Ion product of water

You saw earlier that acids (e.g. hydrochloric acid, HCl; sulphuric acid, H_2SO_4; nitric acid, HNO_3) are compounds which increase the hydrogen ion concentration in aqueous solution, and that these hydrogen ions are hydrated, i.e. exist as H_3O^+. You also saw that bases (e.g. sodium hydroxide, NaOH; potassium hydroxide, KOH; calcium hydroxide, $Ca(OH)_2$), on the other hand, are compounds which increase the hydroxyl ion (OH^-) concentration in aqueous solution.

Water can be regarded as marking the dividing line between acids and bases because it has equal amounts of both of H^+ and OH^- ions in solution. The equilibrium reaction for the dissociation of liquid water is

$$H_2O(l) \rightleftharpoons H^+(aq) + OH^-(aq)$$

with equilibrium constant

$$K_c = \frac{[H^+(aq)][OH^-(aq)]}{[H_2O(l)]}$$

At constant temperature, $[H_2O(l)]$ will remain constant. The equation for the equilibrium constant then becomes:

$$(\text{constant}) \times K_c = [H^+(aq)][OH^-(aq)]$$

The term '(constant) $\times K_c$', which is still a constant, is replaced by the single term K_w, known as the **ion product of water**, so you can write:

$$K_w = [H^+(aq)][OH^-(aq)]$$

The measured value of K_w at 298 K (25°C) is found to be

$$K_w = 1 \times 10^{-14}\,\text{mol}^2\,\text{l}^{-2}$$

Notice, first of all, that the value of this constant limits the concentrations of hydrogen and hydroxyl ions that can coexist in pure water. In water, the $H^+(aq)$ and $OH^-(aq)$ ions are supplied by the dissociation of water. Every molecule of water that dissociates forms one hydrogen and one hydroxyl ion, so if $[H^+]$ is x, $[OH^-]$ must be x as well. From the above,

$$K_w = [H^+][OH^-] = x^2$$
$$= 1 \times 10^{-14}\,mol^2\,l^{-2}$$

Hence
$$x = 1.0 \times 10^{-7}\,mol\,l^{-1}$$

This result shows that in pure water at 298 K (25°C), the hydrogen and hydroxyl ion concentrations are both $1.0 \times 10^{-7}\,mol\,l^{-1}$ (or approximately $2 \times 10^{-6}\,g\,l^{-1}$) — very small values indeed. However, the implications of this equation are rather more wide ranging than this.

Measuring acidity and alkalinity

The equation of K_w is analogous to the solubility product equation; it imposes a limit on the concentration of hydrogen and hydroxyl ions that can coexist in *any* aqueous solution, not just pure water.

For example, consider an aqueous solution of hydrochloric acid (HCl) of concentration $0.1\,mol\,l^{-1}$. In solution, hydrochloric acid is virtually completely dissociated, so that you can write

$$HCl \xrightarrow{H_2O} H^+(aq) + Cl^-(aq)$$

Consequently $0.1\,mol\,l^{-1}$ of HCl will produce a hydrogen ion concentration of $0.1\,mol\,l^{-1}$. But

$$K_w = 1 \times 10^{-14}\,mol^2\,l^{-2} = [H^+][OH^-]$$

Knowing $[H^+]$, you can readily calculate the concentration of OH^- ions; i.e.

$$[OH^-] = \frac{1 \times 10^{-14}\,mol^2\,l^{-2}}{[H^+]}$$
$$= \frac{1 \times 10^{-14}\,mol^2\,l^{-2}}{0.1\,mol\,l^{-1}}$$
$$= 1 \times 10^{-13}\,mol\,l^{-1}$$

Thus the effect of the acid is not only to raise the hydrogen ion concentration from $1 \times 10^{-7}\,mol\,l^{-1}$ in pure water to $0.1\,mol\,l^{-1}$ in the acid, but also to suppress the concentration of OH^- ions from $1 \times 10^{-7}\,mol\,l^{-1}$ in pure water to $1 \times 10^{-13}\,mol\,l^{-1}$ in the acid.

> **Exercise 6.3** The concentration of an aqueous solution of sodium hydroxide (NaOH) is $0.1\,mol\,l^{-1}$. Calculate the concentration of hydrogen ions and hydroxyl ions in this solution. Assume that the sodium hydroxide dissociates completely.

The most noticeable observation from the above calculations is that in moving from a strongly acidic solution to a strongly basic solution, the hydrogen ion concentration has varied over an enormous range (10^{-13} mol l^{-1} to 0.1 mol l^{-1}). You should also note that even in strongly acidic solutions there is still a finite, though small, concentration of OH$^-$ ions, and in strongly basic solutions there is a finite concentration of H$^+$ ions.

How would you define a neutral solution?

A **neutral solution** shows both acidic and basic properties to the same degree. This suggests [H$^+$] = [OH$^-$] as a definition. The value of K_w shows that at 298 K (25°C),

$$[H^+] = [OH^-] = 1.0 \times 10^{-7} \text{ mol l}^{-1}$$

Thus, at 298 K an acidic solution can be defined as one in which [H$^+$] is greater than 10^{-7} mol l^{-1}, and a basic solution is one in which [H$^+$] is less than 10^{-7} mol l^{-1}. This forms the basis of a useful way of measuring the acidity or alkalinity of solutions — ▼The pH scale▲. This scale is widely used in science and technology, and Table 6.8 shows the pH values of some common aqueous solutions.

Strong and weak acids and bases

You've seen that, in aqueous solution, hydrochloric acid dissociates completely, so that

$$\text{HCl} \xrightarrow{H_2O} H^+(aq) + Cl^-(aq)$$

You might expect that an aqueous solution of acetic acid (the principal constituent of vinegar) would dissociate according to

$$\underset{\text{acetic acid}}{CH_3.COOH} \xrightarrow{H_2O} \underset{\text{acetate ion}}{CH_3.COO^-(aq)} + H^+(aq)$$

If therefore you had 0.1 molar solutions of hydrochloric acid and acetic acid, you might also expect that the solutions will have the same pH values because they both have the same number of H$^+$ ions.

However, look at Table 6.8. A 0.1 molar solution of hydrochloric acid has a pH of 1.0, as expected from the earlier calculations, but a 0.1 molar solution of acetic acid has a pH of 2.9. In other words, the acetic acid is less 'acidic' than the hydrochloric acid. Remember that, earlier, an acid was defined as a compound which increases the hydrogen ion concentration in solution, and it does this by dissociating in solution into ions. Since hydrochloric acid and acetic acid both produce the same number of ions when they dissociate, the only conclusion you can draw from the different pH values in Table 6.8 is that some of the dissolved acetic acid has not dissociated.

Table 6.8 pH values of some common solutions

Solution	pH
hydrochloric acid (0.1 mol l^{-1})	1.0
gastric juice (human)	1.0–2.5
lemon juice	≈2.1
acetic acid (0.1 mol l^{-1})*	2.9
orange juice	≈3.0
tomato juice	≈4.1
rainwater (depending on local conditions)	≈5.0
urine	6.0
saliva (human)	6.8
milk	≈6.9
pure water (25°C)	7.0
blood (human)	7.4
sea water	7.9–8.3
ammonia (0.1 mol l^{-1})**	11.1
sodium hydroxide (0.1 mol l^{-1})	13.0

*Approximates to household vinegar
**Approximates to household solutions

▼The pH scale▲

Let's do an imaginary experiment. Suppose you have one litre of hydrochloric acid solution of concentration 0.1 mol l^{-1}; the concentrations of H$^+$(aq) and OH$^-$(aq) will be those calculated earlier: [H$^+$] = 0.1 mol l^{-1}, and [OH$^-$] = 1.0 × 10^{-13} mol l^{-1}. Suppose that you now slowly add 0.1 mol of solid sodium hydroxide. A neutralization reaction occurs: H$^+$(aq) is consumed and the 'acidity' of the solution drops. When all of the solid has been added, you end up with a solution of sodium chloride in water. Now the concentrations of H$^+$(aq) and OH$^-$(aq) will be those for pure water; that is, [H$^+$] = [OH$^-$] = 1.0 × 10^{-7} mol l^{-1}.

Thus, by adding NaOH to the HCl solution, the concentration of H$^+$(aq) has decreased by a factor of 10^6 (a million) from 10^{-1} to 10^{-7} mol l^{-1}; at the same time, the hydroxyl ion concentration has risen by the same factor.

Suppose that you wanted to show graphically how the acidity (that is, concentration of H$^+$(aq)) changes during the course of this experiment. How would you even choose a scale for such a graph when the quantity of interest changes by six orders of magnitude? Fortunately, there is a way out. The concentration of H$^+$(aq) drops by a million from 10^{-1} to 10^{-7} mol l^{-1}, so that the index drops from −1 to −7. It looks as if this index might provide a much more manageable measure of concentration and concentration changes when they cover many orders of magnitude. This index is, of course, the logarithm to the base ten of the number in question, or log$_{10}$. (Remember 'Logarithms and exponentials' on p. 41 of Unit 3.)

Table 6.9 shows values of [H$^+$], the corresponding [OH$^-$], and log$_{10}$[H$^+$] within the range most commonly encountered in chemistry. As you can see, almost all values of log$_{10}$[H$^+$] are negative numbers. The **pH scale** (meaning 'potential of hydrogen') is therefore based on −log$_{10}$[H$^+$], with the pH value of a solution defined by

$$\mathrm{pH} = -\log_{10}[\mathrm{H}^+] \qquad (6.9)$$

Table 6.9

[H$^+$] mol l^{-1}	[OH$^-$] mol l^{-1}	log$_{10}$[H$^+$]	pH
10^1 or 10	10^{-15}	1	−1
10^0 or 1	10^{-14}	0	0
10^{-1} or 0.1	10^{-13}	−1	1
10^{-2} or 0.01	10^{-12}	−2	2
10^{-3} or 0.001	10^{-11}	−3	3
10^{-4} or 0.000 1	10^{-10}	−4	4
10^{-5} or 0.000 01	10^{-9}	−5	5
10^{-6} or 0.000 001	10^{-8}	−6	6
10^{-7} or 0.000 000 1	10^{-7}	−7	7
10^{-8} or 0.000 000 01	10^{-6}	−8	8
10^{-9} or 0.000 000 001	10^{-5}	−9	9
10^{-10} or 0.000 000 000 1	10^{-4}	−10	10
10^{-11} or 0.000 000 000 01	10^{-3}	−11	11
10^{-12} or 0.000 000 000 001	10^{-2}	−12	12
10^{-13} or 0.000 000 000 000 1	10^{-1}	−13	13
10^{-14} or 0.000 000 000 000 01	10^0	−14	14

The resulting values of pH are shown in the final column of Table 6.9. So pure water at 298 K (25°C), in which the concentration of hydrogen ions is 1 × 10^{-7} mol l^{-1}, has a pH of 7. Acidic solutions, in which the hydrogen ion concentration is greater than 10^{-7} mol l^{-1}, will have a pH of less than 7 and basic solutions, in which the hydrogen ion concentration is less than 10^{-7} mol l^{-1}, will have a pH greater than 7.

Thus, the dissociation equation is better written as

$$\mathrm{CH_3.COOH} \underset{}{\overset{H_2O}{\rightleftharpoons}} \mathrm{CH_3.COO^-(aq) + H^+(aq)}$$

Acids which completely dissociate in solution are called **strong acids**, and solutions of such acids contain only ions. In contrast, solutions, such as acetic acid, which only partially dissociate in solution, are called **weak acids**. Such solutions contain a mixture of ions and undissociated molecules.

The same type of argument can be used for bases. Sodium hydroxide in aqueous solution dissociates completely as

$$\mathrm{NaOH(s)} \xrightarrow{H_2O} \mathrm{Na^+(aq) + OH^-(aq)}$$

so it is a **strong base**. Ammonia, on the other hand, only partially dissociates in solution

$$NH_3(aq) + H_2O(l) \rightleftharpoons NH_4^+(aq) + OH^-(aq)$$

so it is a **weak base**. This is borne out by Table 6.8, where you can see that the pH of a 0.1 molar solution of sodium hydroxide is 13, whereas that of an equivalent ammonia solution is only 11.1.

As you've seen, calculating the hydrogen ion concentration in solutions of strong acids and strong bases, where complete dissociation has occurred, is straightforward. To do the same thing for weak acids and bases, you need some information about the proportion of the dissolved compound which has dissociated.

For a weak acid (general formula, HA), the acid molecules in solution, HA(aq) must be in equilibrium with the ions in solution, and so

$$HA(aq) \xrightleftharpoons[]{H_2O} H^+(aq) + A^-(aq)$$

The concentration equilibrium constant K_c for this reaction is

$$K_c = \frac{[H^+(aq)][A^-(aq)]}{[HA(aq)]}$$

$[H^+(aq)]$ and $[A^-(aq)]$ are the concentrations of the ions in the solution and $[HA(aq)]$ is the concentration of undissociated acid in solution.

When applied to the dissociation of acids in solution, this type of equilibrium constant is known as the **acid dissociation constant**, symbol K_a, so that

$$K_a = \frac{[H^+(aq)][A^-(aq)]}{[HA(aq)]} \text{ or } \frac{[H^+][A^-]}{[HA]}$$

Typical values of K_a for selected acids are shown in Table 6.10. (▼Atmospheric acids▲ on p.104 explores one area where the strengths of acids are relevant.) There are three important points about Table 6.10.

(a) Dissociation constants are not used for strong acids. As noted earlier, these acids are completely dissociated in aqueous solution so that $[HA(aq)]$ would be zero and K_a would consequently be infinitely large.

(b) K_a measures the strength of an acid. Thus in Table 6.10, iodic acid is stronger than nitrous acid because is has a higher value of K_a. The strengths of strong acids are never compared; nitric acid and hydrochloric acid are equally strong since they are both completely dissociated in solution.

(c) Notice that carbonic acid, H_2CO_3, dissociates by a two-stage process, each stage having its own dissociation constant. This multiple-stage dissociation occurs for most weak acids which contain more than one hydrogen atom. Phosphoric acid, H_3PO_4, for example, dissociates in three stages. In general, when multi-stage dissociation occurs, the dissociation constants become progressively smaller.

Table 6.10   Dissociation of weak acids

Acid	Dissociation equation	$\dfrac{K_a}{\text{mol l}^{-1}}$
iodic	$HIO_3(aq) \rightleftharpoons H^+(aq) + IO_3^-(aq)$	1.76×10^{-1}
nitrous	$HNO_2(aq) \rightleftharpoons H^+(aq) + NO_2^-(aq)$	4.50×10^{-4}
acetic	$CH_3.COOH(aq) \rightleftharpoons H^+(aq) + CH_3.COO^-(aq)$	1.78×10^{-5}
carbonic	$H_2CO_3(aq) \rightleftharpoons H^+(aq) + HCO_3^-(aq)$	4.20×10^{-7}
hydrogen carbonate	$HCO_3^-(aq) \rightleftharpoons H^+(aq) + CO_3^{2-}(aq)$	4.80×10^{-11}

Summary

- The equilibrium point in an equilibrium reaction occurs when the rates of reaction in both directions are the same.
- The equilibrium constant, K, is defined by

$$K = \frac{\text{(product of concentrations of products)}}{\text{(product of concentrations of reactants)}}$$

- The yield of a reaction is given by

$$\text{yield} = \frac{\text{(actual concentration of a product)}}{\text{(expected concentration in completed reaction)}}$$

- The solubility product, K_{sp}, for an ionic compound AB is defined by

$$K_{sp} = [A^+][B^-]$$

- The ion product, K_w, of water is defined as

$$K_w = [H^+][OH^-]$$

- The pH value of a solution is defined by

$$\text{pH} = -\log_{10}[H^+]$$

- Strong acids and bases dissociate completely; weak ones only partially. The strength of a weak acid or base is described by the appropriate equilibrium dissociation constant.

SCIENCE FOR MATERIALS

▼Atmospheric acids▲

Rainwater is naturally acidic, with an average pH of about 5.0 (Table 6.9), the exact value depending on local conditions, weather etc.

So where does this acidity arise, if neutral water has a pH of 7? There are at least three natural sources of acid gas: carbon dioxide in the atmosphere, nitrogen oxides (NO_x) produced by lightning discharges, and sulphur oxides (SO_2, SO_3) from volcanic action. Carbonic acid is formed by dissolution of CO_2:

$$CO_2(g) + H_2O(l) \rightleftharpoons H_2CO_3(aq)$$

followed by the dissociation reactions shown in Table 6.10. Carbonic acid is very weak, and naturally leads to a pH of about 5.6. NO_x gases, of which NO_2 is the most important, are formed in the very high temperature zone surrounding forked lightning:

$$N_2(g) + O_2(g) \rightleftharpoons 2NO(g)$$
$$2NO(g) + O_2(g) \longrightarrow 2NO_2(g)$$

Further oxidation occurs to produce nitric acid:

$$NO_2(g) + O_2(g) + 2H_2O(l) \longrightarrow 4HNO_3(aq)$$

Sulphuric acid is formed by oxidation of SO_2 present in volcanic clouds:

$$2SO_2(g) + O_2(g) \rightleftharpoons 2SO_3(g)$$
$$SO_3(g) + H_2O(l) \longrightarrow H_2SO_4(aq)$$

Although both nitric and sulphuric acids are very strong, their concentration is much lower than carbonic acid, so the pH falls only to about 5 when they're present. Human pollution can increase NO_x (car exhausts mainly), while SO_2 levels are large downwind of power stations where sulphur-rich coal or oil are burnt (Figure 6.16). Such polluted rain can have a pH as low as 4.1 (Central Europe), whilst thunderstorms can produce pH 3 rain.

The deleterious effects of acid rain are serious, particularly for old limestone structures (e.g. Figure 6.17), where the general reaction occurs:

$$CaCO_3(s) + H_3O^+(aq) \longrightarrow Ca^{2+}(aq) + HCO_3^-(aq) + H_2O$$

Steel structures are also susceptible (as the regular repainting of structures like the Forth Rail Bridge bears witness):

$$Fe(s) + H_2SO_4(aq) \longrightarrow FeSO_4(aq) + H_2(g)$$

The costs of acid rain are thus very high. So what can be done to ameliorate such pollution? Since coal-burning power stations are a primary source of SO_2, flue–gas desulphurization offers one way of tackling the problem. The furnace gases are washed with a limestone or chalk ($CaCO_3$) slurry:

$$SO_2(g) + H_2O(l) \longrightarrow H_2SO_3(aq)$$
$$CaCO_3(s) + H_2SO_3(aq) \longrightarrow CaSO_3(aq) + CO_2(g) + H_2O(l)$$
$$2CaSO_3(aq) + O_2(g) \longrightarrow 2CaSO_4(s)$$
$$CaSO_4 + 2H_2O \longrightarrow CaSO_4.2H_2O(s)$$

The final product is gypsum, which can be processed further to make the building material, plaster board. Notice that CO_2 gas is produced in the reaction sequence, so one seriously acidic pollutant is replaced by a weakly acidic one. Whatever the success with such projects, acid rain will remain with us as a problem for architects, designers and materials engineers to overcome!

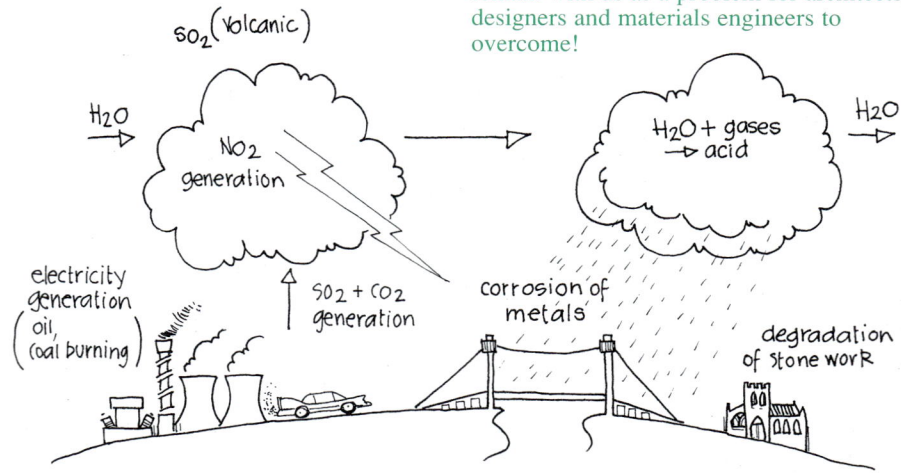

Figure 6.16 Atmospheric acids associated with the water cycle

Figure 6.17 The Rollright Stones, Oxfordshire, showing the effects of 4000 years of acid rain

6.4 Reaction kinetics

The second half of Part 1 *Chemistry* of Videocassette 2 and its associated notes explore reaction kinetics. You should study them in conjunction with this section.

When natural gas, predominantly methane (CH_4), burns in air, it combines with oxygen to produce carbon dioxide (CO_2) and water (H_2O). You've already seen that the balanced chemical equation for this reaction is

$$CH_4(g) + 2O_2(g) \longrightarrow CO_2(g) + 2H_2O(g)$$

If you're told that the reaction has an equilibrium constant of 10^{40} at 298 K (25°C), you should realise that, when the reaction occurs, practically all of the reactants (CH_4 and O_2) will be converted to products (CO_2 and H_2O).

But if you've ever turned a gas tap on, you'll know that, although gas comes out, there's no reaction until you put a light to it. Why is this so when equilibrium considerations tell you that the reactants should be almost totally converted to products? The answer lies in the **reaction rate**; that is, the speed with which the products are formed. For methane reacting with oxygen at 298 K, the reaction rate is so slow as to be imperceptible.

Thus, if you want to describe a reaction, and especially if you want to harness the reaction in some production process, you need two vital pieces of information. First, you need to know the yield of products. This information comes from examining the equilibrium constant for the reaction as you saw in the previous section. Secondly, you need to know how fast the reaction will proceed; i.e. how long will it take for the equilibrium concentration of products to be formed? This information comes from examining reaction rate — the subject of this section.

6.4.1 Defining reaction rate

To describe the rate of reaction you need some variable of the reaction which changes with time. To illustrate how this can be done, let's examine the decomposition of nitrogen pentoxide (N_2O_5) into nitrogen peroxide (N_2O_4) and oxygen. The overall reaction can be represented by:

$$N_2O_5(g) \longrightarrow N_2O_4(g) + \frac{1}{2}O_2(g)$$

If a sample of nitrogen pentoxide is placed in a reaction vessel and *held at constant temperature and volume*, there will initially be only nitrogen pentoxide present. However, as the decomposition reaction proceeds, the concentration of N_2O_5 will decrease and the concentrations of N_2O_4 and O_2 will increase.

You could follow the progress of this reaction by measuring the concentration of N_2O_5 at regular time intervals. If, in a small time interval Δt, the concentration of N_2O_5 changes by an amount $\Delta[N_2O_5]$, the rate of reaction can be written as

$$\text{rate} = \frac{\Delta[N_2O_5]}{\Delta t}$$

There is nothing particularly special about monitoring $[N_2O_5]$. You could equally well have measured the reaction rate by monitoring the change in concentration of N_2O_4 or O_2 with time, i.e.

$$\frac{\Delta[N_2O_4]}{\Delta t} \text{ or } \frac{\Delta[O_2]}{\Delta t}$$

However, all of these rates are interdependent. The rate at which N_2O_4 appears is exactly the same as the rate at which N_2O_5 disappears since each molecule of N_2O_5 produces one molecule of N_2O_4. So

$$-\frac{\Delta[N_2O_5]}{\Delta t} = \frac{\Delta[N_2O_4]}{\Delta t}$$

where the minus sign indicates that N_2O_5 is *disappearing* whereas N_2O_4 is *appearing* and so is positive.

Similarly, the reaction equation shows that for every molecule of N_2O_5 that disappears only half a molecule of O_2 is produced. Or, more realistically, two molecules of N_2O_5 have to disappear before one molecule of O_2 appears. Consequently

$$-\frac{1}{2}\frac{\Delta[N_2O_5]}{\Delta t} = \frac{\Delta[O_2]}{\Delta t}$$

When the time interval Δt becomes infinitesimally small, the reaction rate can be written as the derivative $(d[X]/dt)$, where $[X]$ is the concentration of one of the participants (measured typically in mol l^{-1}) and t is time measured in seconds. In general, then, for a reaction of the form

$$A \longrightarrow B + C$$

(e.g. the HI reaction you saw earlier) carried out at constant temperature and volume, the rate of reaction can be described by any of the terms $(d[A]/dt)$ or $(d[B]/dt)$ or $(d[C]/dt)$. Note that for a reaction of the form $A \longrightarrow B + C$,

$$-\frac{d[A]}{dt} = \frac{d[B]}{dt} = \frac{d[C]}{dt}$$

6.4.2 Variation of concentration with time

To see how $(d[X]/dt)$ might be evaluated, consider again the simple hypothetical reaction in which a compound A dissociates into two new compounds B and C at constant temperature. As the reaction proceeds, [A] will decrease in time, and Figure 6.18 shows the kind of decrease that you might measure. Such graphs are called **reaction profiles**.

The reaction rate also decreases as the reaction proceeds, so it may be dependent in some way on the concentration of A. For example, it might be that the reaction rate is directly proportional to [A], in which case you could write

$$-\frac{d[A]}{dt} \propto [A]$$

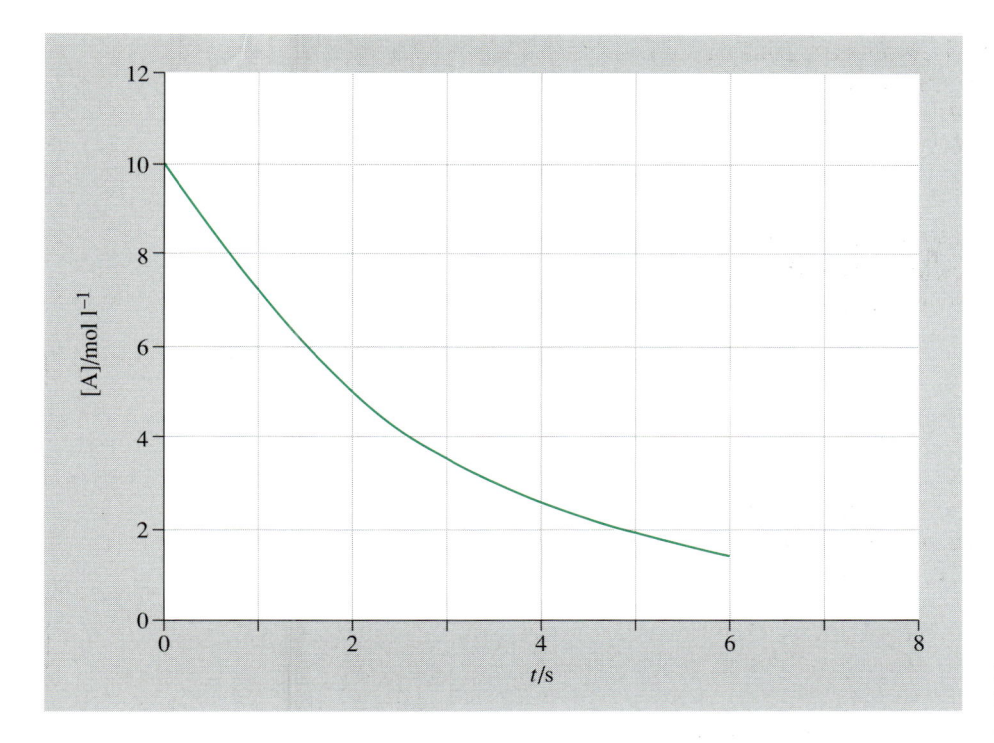

Figure 6.18 Plot of [A] as a function of time

(minus because the rate is decreasing), or

$$-\frac{d[A]}{dt} = k[A] \tag{6.10}$$

where k is a constant known as the **rate constant**. It turns out that Equation (6.10) applies to the dissociation reaction of N_2O_5 gas, so that

$$-\frac{d[N_2O_5]}{dt} = k[N_2O_5]$$

The reaction is known as a **first-order reaction** since its rate depends on the concentration of N_2O_5 to the first power.

6.4.3 First-order reactions

Equation (6.10) for a first order reaction can be rearranged by 'separating the variables' to give

$$-\frac{1}{[A]} d[A] = k\, dt$$

and this can be integrated between specified limits.

If initially at $t = 0$, $[A] = [A]_0$, you can write the integration as

$$-\int_{[A]_0}^{[A]} \frac{1}{[A]} d[A] = \int_0^t k\, dt$$

Integrating this gives

$$-(\ln[A] - \ln[A]_0) = kt$$

or

$$\ln[A] - \ln[A]_0 = -kt \qquad (6.11)$$

This result arises because, as you saw in Unit 3,

$$\int \frac{1}{[A]} d[A] = \ln [A]$$

(note, natural logarithms).

Rearranging the terms in Equation (6.11) gives

$$\ln[A] = -kt + \ln[A]_0. \qquad (6.12)$$

But the general equation of a straight line is of the form

$$y = mx + c$$

where m is the slope of the line and c is the intercept of the line on the y axis (where $x = 0$). Equation (6.12) is thus the equation of a straight-line graph of $\ln[A]$ versus t, whose slope is $-k$ and whose intercept on the $\ln[A]$ axis is $\ln[A]_0$. This is shown schematically in Figure 6.19.

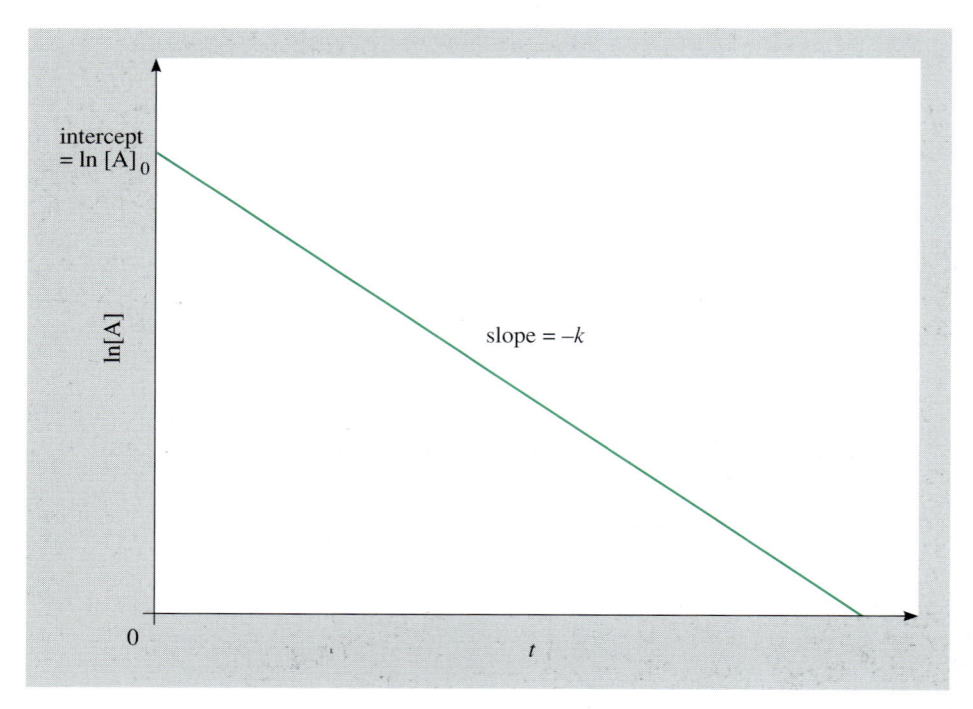

Figure 6.19 Schematic plot of $\ln[A]$ versus t

Here, then, is a very simple test to determine whether or not a reaction is first order. Plot the results of the logarithm of concentration versus time, and if a straight-line graph is produced the reaction is first order. It's also a simple matter to determine the rate constant k from the slope of the graph.

To illustrate this, the first two columns of Table 6.11 show how [A] varies with time for the hypothetical reaction depicted in Figure 6.18. The data are plotted in Figure 6.20 as ln[A] versus t. You can see that the results lie close to a straight line whose slope works out as $-0.35\,\text{s}^{-1}$. The fact that the graph is a straight line demonstrates that the reaction is first order and from the slope, its rate constant, $k = 0.35\,\text{s}^{-1}$.

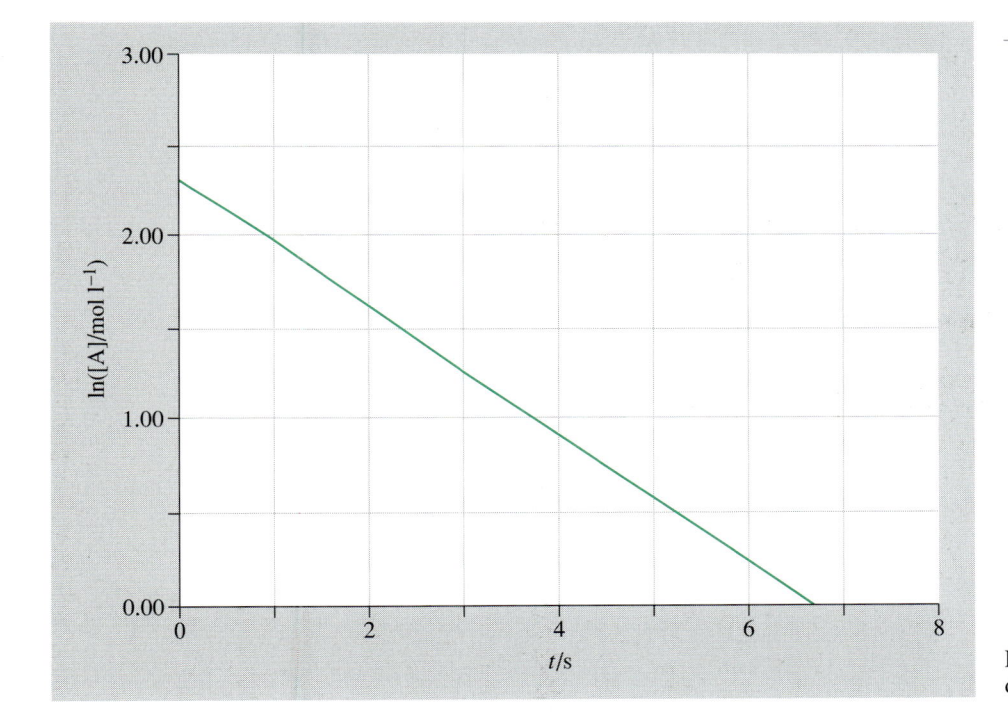

Table 6.11 Variation of [A] with time

$\dfrac{t}{\text{s}}$	$\dfrac{[A]}{\text{mol l}^{-1}}$	ln[A]
0	10.0	2.30
1	7.2	1.97
2	5.0	1.61
3	3.5	1.25
4	2.5	0.92
5	1.8	0.59
6	1.3	0.26

Figure 6.20 Plot of ln[A] versus t for the data in Table 6.11

SAQ 6.14 (Objectives 6.9 and 6.10)
Nitrogen pentoxide (N_2O_5) dissociates at 310 K according to the equation

$$N_2O_5(g) \longrightarrow N_2O_4(g) + \tfrac{1}{2}O_2(g)$$

and Table 6.12 gives experimentally measured data for the concentration of N_2O_5 (measured as the partial pressure of the gas) as a function of time. Use these data to confirm that it is a first-order reaction. Determine the rate constant for the reaction.

Table 6.12 Data for SAQ 6.14

$\dfrac{t}{\text{s}}$	$\dfrac{[N_2O_5]}{\text{kN m}^{-2}}$
1 200	33.5
3 000	26.6
6 000	18.1
8 400	13.3
12 000	8.4

SCIENCE FOR MATERIALS

Pseudo first–order reactions

The number of reactions which are first order involving a single reactant is relatively small. There is, however, a significant number of reactions involving more than one reactant which can nevertheless be analysed by treating them as first order with respect to one of the reactants. To illustrate this, consider the hydrolysis reaction:

$$CH_3.COO.CH_3 + H_2O \longrightarrow CH_3.COOH + CH_3.OH$$

methyl acetate $\qquad\qquad$ acetic acid + methanol

This equation suggests that the rate at which the reactants disappear should depend both on the amount of acetate and the amount of water present, since both are involved in the reaction. So we would expect the reaction to be of a higher order than first order. If, however, the reaction is monitored by following changes in the concentration of methyl acetate, typical results are as shown in Table 6.13. When these results are plotted as $\ln[CH_3.COO.CH_3]$ versus time they approximate closely to a straight line, as shown in Figure 6.21 (opposite). This shows that the reaction is effectively first order with respect to methyl acetate even though the overall chemical equation suggests that, with water involved, the reaction should be of a higher order.

Table 6.13 Hydrolysis of methyl acetate

$\dfrac{t}{s}$	$[CH_3.COO.CH_3]$ $mol\,l^{-1}$	$\ln[CH_3.COO.CH_3]$
0	0.308	−1.18
1 200	0.288	−1.25
4 500	0.241	−1.42
7 140	0.209	−1.57

To understand this apparent anomaly, you need to examine the details of the reaction. When the methyl acetate is dissolved in water, Table 6.13 shows that the initial concentration of methyl acetate is only 0.308 mol l^{-1}. This is equivalent to 22.8 ml per litre of solution. In other words, there is a large excess of water present. The change in the concentration of water in the mixture, even when the reaction has gone to completion, is negligible compared with the change in concentration of the acetate. So the reaction can be treated as first order with respect to the methyl acetate. Such reactions are known as **pseudo first–order**.

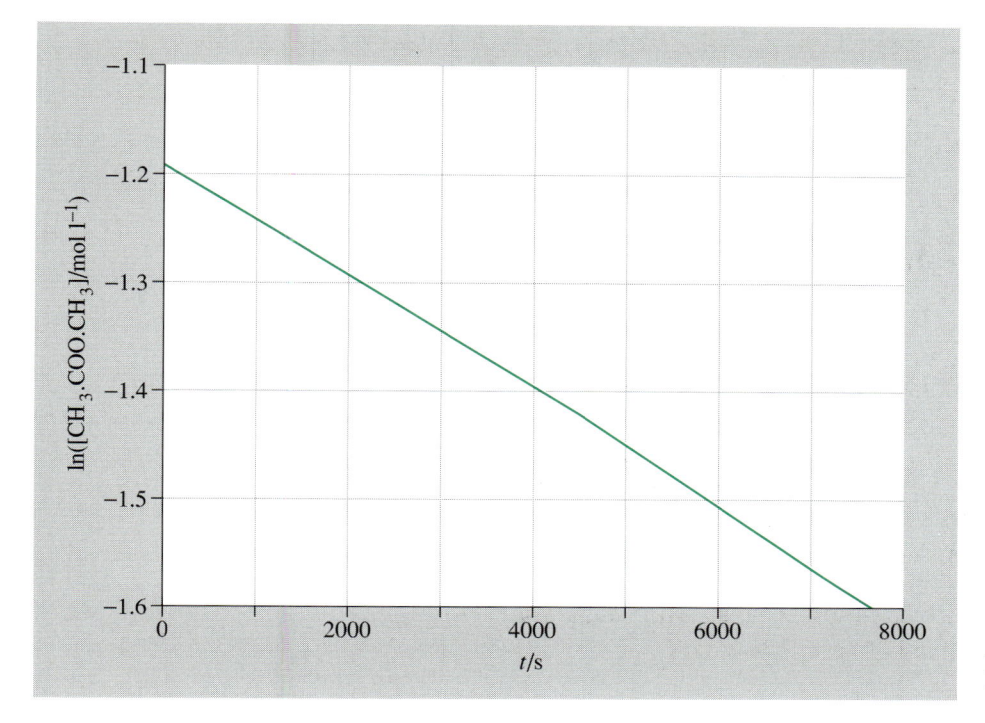

Figure 6.21 Plot of ln[CH$_3$.COO.CH$_3$] versus time from the data in Table 6.13

SAQ 6.15 (Objectives 6.9 and 6.10)
Using either the numerical data shown in Table 6.13 or the graphical information of Figure 6.21, calculate the pseudo first-order rate constant for the hydrolysis of methyl acetate at 298 K. What will be the concentration after 4 hours of:
(a) methyl acetate?
(b) acetic acid?

6.4.4 Effect of temperature on reaction rate

It has long been known that chemical reactions speed up when the temperature is raised. So when rate constants are quoted they should always specify the temperature at which they were measured.

To show how rate constant varies with temperature, Table 6.14 gives values of rate constant at different temperatures for the decomposition of nitrogen pentoxide that you considered earlier

$$N_2O_5(g) \longrightarrow N_2O_4(g) + \frac{1}{2}O_2(g)$$

The variation is shown graphically in Figure 6.22 (overleaf). You can see that the rate constant, and thus the reaction rate, increase increasingly rapidly as the temperature is raised. Why is this?

Table 6.14 Decomposition of nitrogen pentoxide

$\dfrac{T}{K}$	$\dfrac{k}{(Ms)^{-1}}$
273	0.008
298	0.346
308	1.35
318	4.98
328	15.0
338	48.7

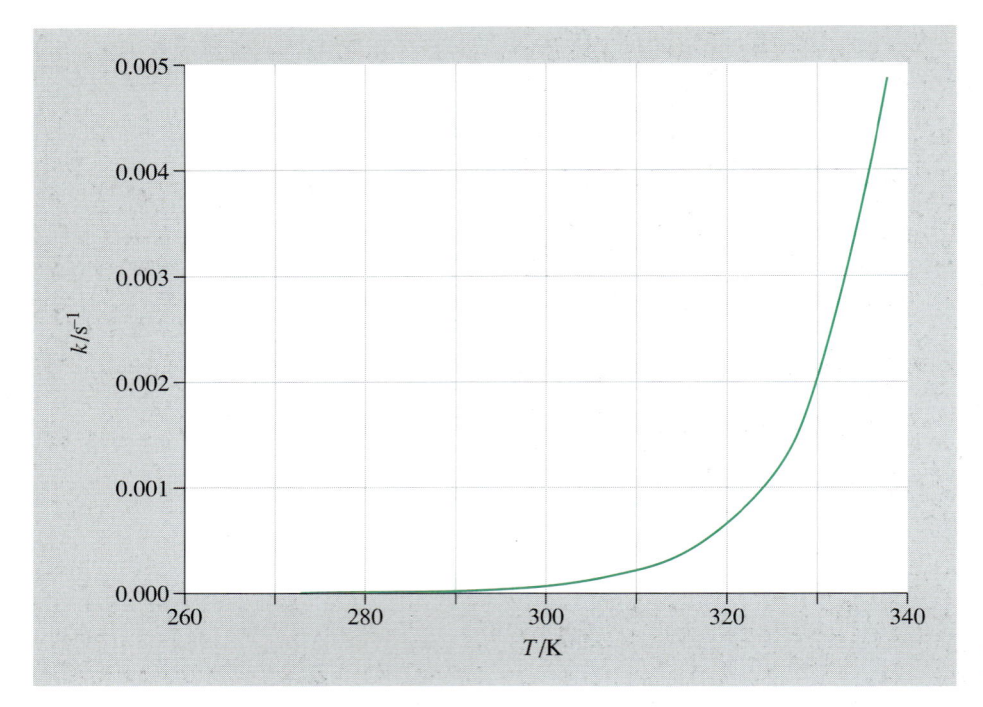

Figure 6.22 Rate constant k versus temperature T for the decomposition of nitrogen pentoxide (Table 6.14)

In order for a molecule to dissociate (first-order reaction), or for two or more molecules to react together (higher order reactions), they have to pass through an intermediate, higher energy state. This extra energy required is known as the **activation energy** for the reaction. When two molecules come together and react (as in the hydrolysis of methyl acetate or dissociation of hydrogen iodide), this is the energy of the two molecules when they meet. This higher energy combination is known as an **activated complex** (Figure 6.23, opposite), and much of its energy comes from the kinetic energy of thermal motion.

Two examples of activated complexes are shown in Figure 6.24. For the dissociation of N_2O_5, the activated complex is just a single molecule. Translational kinetic energy cannot be the source of the activation energy here. In this case, it comes from the vibrational energy of the bonds between the atoms in the molecule. In N_2O_5, the weakest bond is the axial N–O–N bond. Given enough vibrational energy, it will break first and dissociate to give two NO_2 fragments (which promptly recombine to form N_2O_4) and an activated oxygen atom. This also combines rapidly with another lone oxygen atom to form an oxygen molecule (Figure 6.24a).

In the case of hydrogen iodide, the activated complex is two molecules aligned side-by-side. With enough thermal energy, the electron bonds will swap (the smaller arrows in Figure 6.24(b), to give hydrogen and iodine gas.

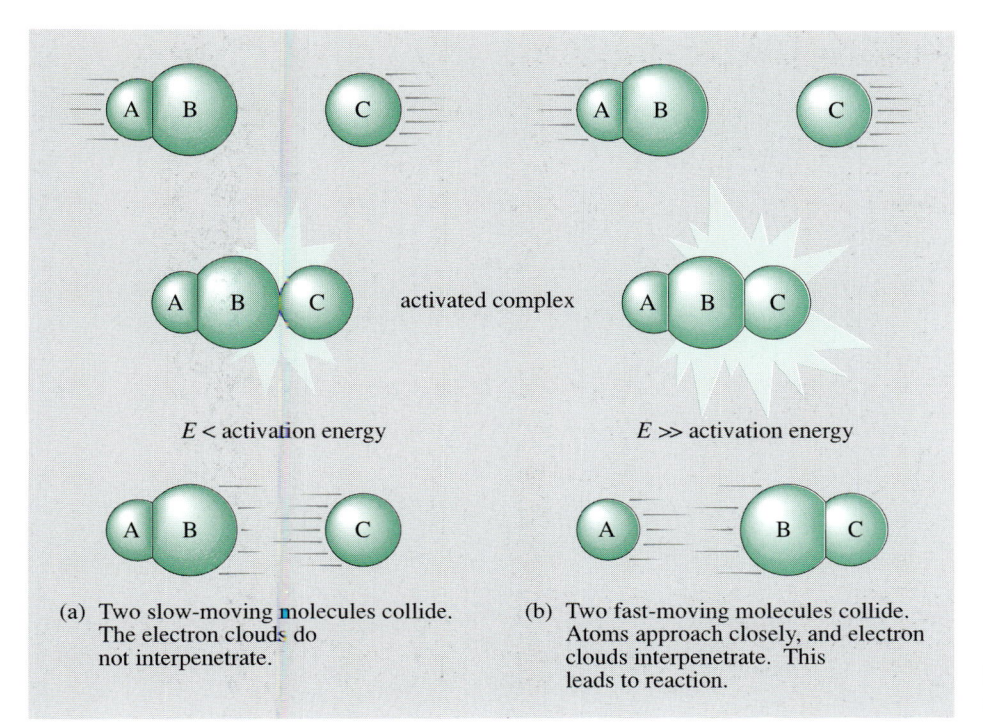

Figure 6.23 Molecular interpretation of activation energy

(a) Two slow-moving molecules collide. The electron clouds do not interpenetrate.

(b) Two fast-moving molecules collide. Atoms approach closely, and electron clouds interpenetrate. This leads to reaction.

Figure 6.24 Activated complexes of (a) nitrogen pentoxide, and (b) hydrogen iodide

In effect, the activation energy represents an energy barrier to be overcome for the reaction to occur, as shown schematically in Figure 6.25 (overleaf). Here $E_{A(F)}$ represents the activation energy of the forward reaction and $E_{A(R)}$ is the activation energy of the reverse reaction. The difference between the two is just the overall enthalpy of reaction, $(-\Delta H_R^\ominus)$ if performed under standard conditions. As the temperature is raised, a greater proportion of the molecules acquire sufficient energy to overcome this energy barrier, and thus the reaction rate increases. You'll be looking at this in more detail in the next Block of the course.

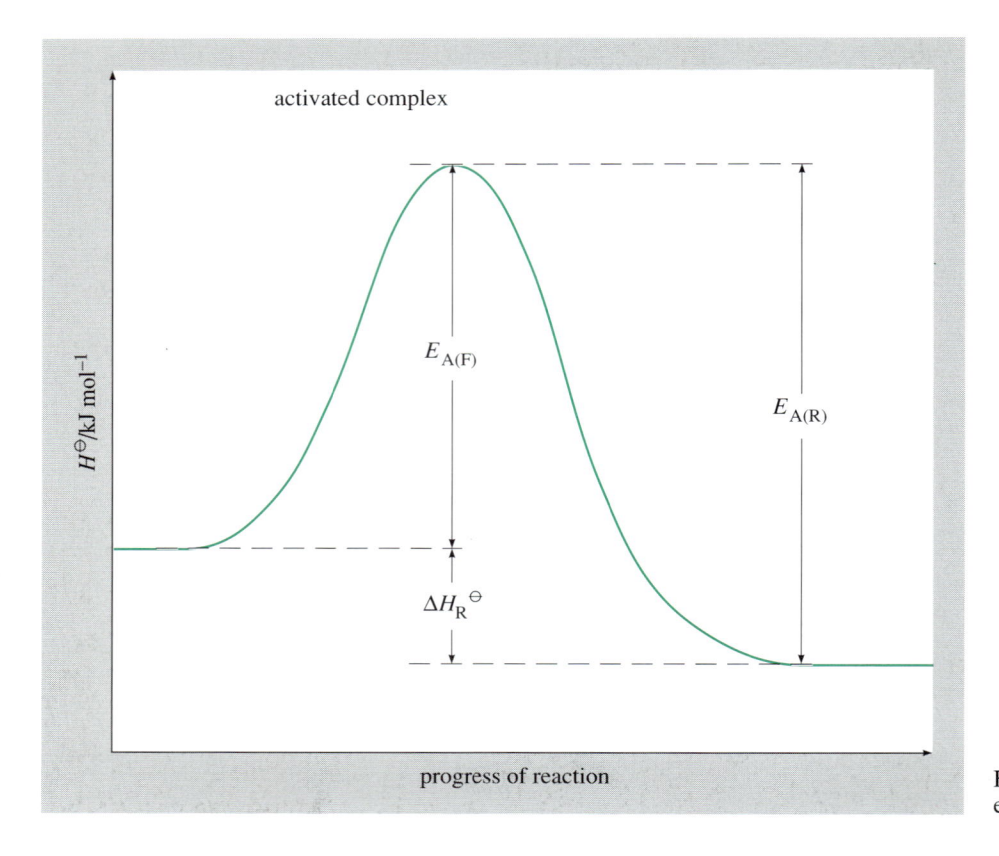

Figure 6.25 Schematic of activation energies for forward and reverse reactions

Catalysis

There is another effect that complicates reaction kinetics. Methane and oxygen will not normally react together at room temperature, but need something like a high temperature spark to initiate the reaction. However, they will react together even at room temperature if a **catalyst** is present. For example, if a fine platinum or palladium gauze is inserted into a gaseous mixture of methane and oxygen, it will heat up and eventually glow red-hot as it catalyses the oxidation reaction. When all the gases have reacted, the gauze will cool back to room temperature and is totally unaffected by its recent, albeit rather violent, experience.

A catalyst effectively speeds up a reaction by promoting the formation of an activated complex, but is itself unaffected by the process. The effect is illustrated schematically in Figure 6.26 (opposite). So what is happening at a molecular level?

It's thought that the surfaces of many metals, especially transition metals, have active surfaces onto which molecules are attracted, form an activated complex and react together (Figure 6.27). There is an element of positive feedback here, because as the reaction occurs, heat is liberated (ΔH is large and negative for all combustion reactions) and heats up the catalyst. Since it now has greater thermal energy (of vibration of the atoms in the crystal lattice), more energy is available for the activated complex.

Figure 6.26   The energetics of catalysis

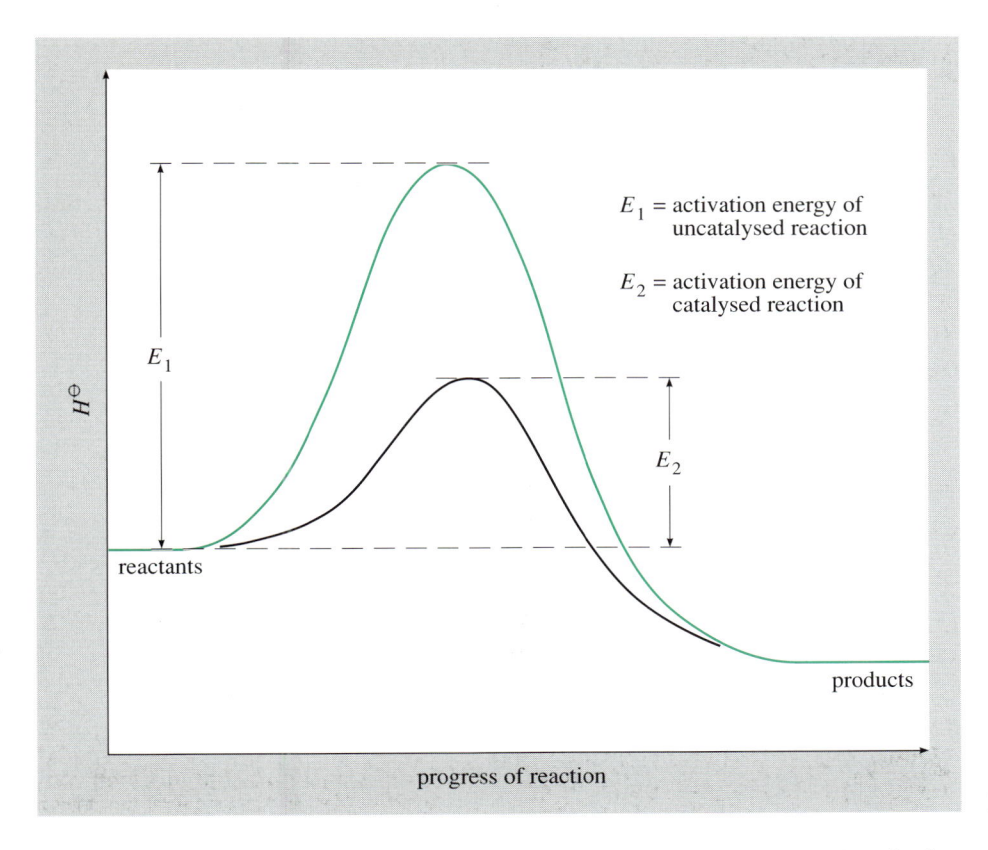

Such catalysts are of course, widely exploited in many large-scale chemical processes because they reduce the energy costs, enabling many reactions to be conducted at much lower temperatures than if they were uncatalysed. Such processes include the fixation of nitrogen to form ammonia (for fertilizers), the cracking of hydrocarbons to form ethylene etc (catalytic cracking), as well as many polymerization reactions (e.g. to form high density polyethylene). A more familiar example, however, is the ▼Catalytic converter▲ (p.116) fitted to a motor car exhaust.

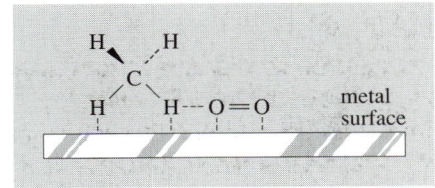

Figure 6.27   Molecular interpretation of catalysis during reaction of methane with oxygen

Summary

- The rate of a reaction is described by the rate of change of concentration of one of its constituents with time.

- The rate constant of a reaction is the constant of proportionality between the reaction rate and the concentration of one of its constituents raised to some power — in first order reactions, the power is one; in second order, two, and so on.

- Rate constants are strongly temperature-dependent. As temperature increases, more of the reacting molecules acquire kinetic energy greater than the activation energy for the intermediate activated complex.

- Catalysts promote the formation of activated complexes, in effect lowering the activation energy. They increase the rate of a reaction without themselves being changed.

SCIENCE FOR MATERIALS

▼Catalytic converter▲

Since April 1993, all new cars in the UK have had to be fitted with a catalytic converter. The device is fitted to the exhaust pipe (Figure 6.28(a)) and consists of a stainless steel can into which is fitted a ceramic honeycomb (Figures 6.28(b) and (c)). The device helps to reduce pollution from unburnt hydrocarbons (HC) from the petrol, NO_x produced by nitrogen oxidation in the combustion chambers and carbon monoxide (CO) from incomplete combustion in the engine. Not only do these contribute to acid rain (see 'Atmospheric acids'), but they can also react further in bright sunlight to give toxic ozone gas (the main ingredient in so-called photochemical smog).

The active components of the catalytic converter are finely divided platinum and rhodium present in a matrix of alumina on the surfaces of the cells in the honeycomb. The platinum metal oxidizes the hydrocarbons and CO to CO_2 and water using the oxygen produced by the reactions catalysed by the rhodium, where the NO_x is reduced to nitrogen and oxygen gases. Such a 'three-way converter' is designed to operate at a temperature of about 600°C, so it is relatively inefficient during start-up.

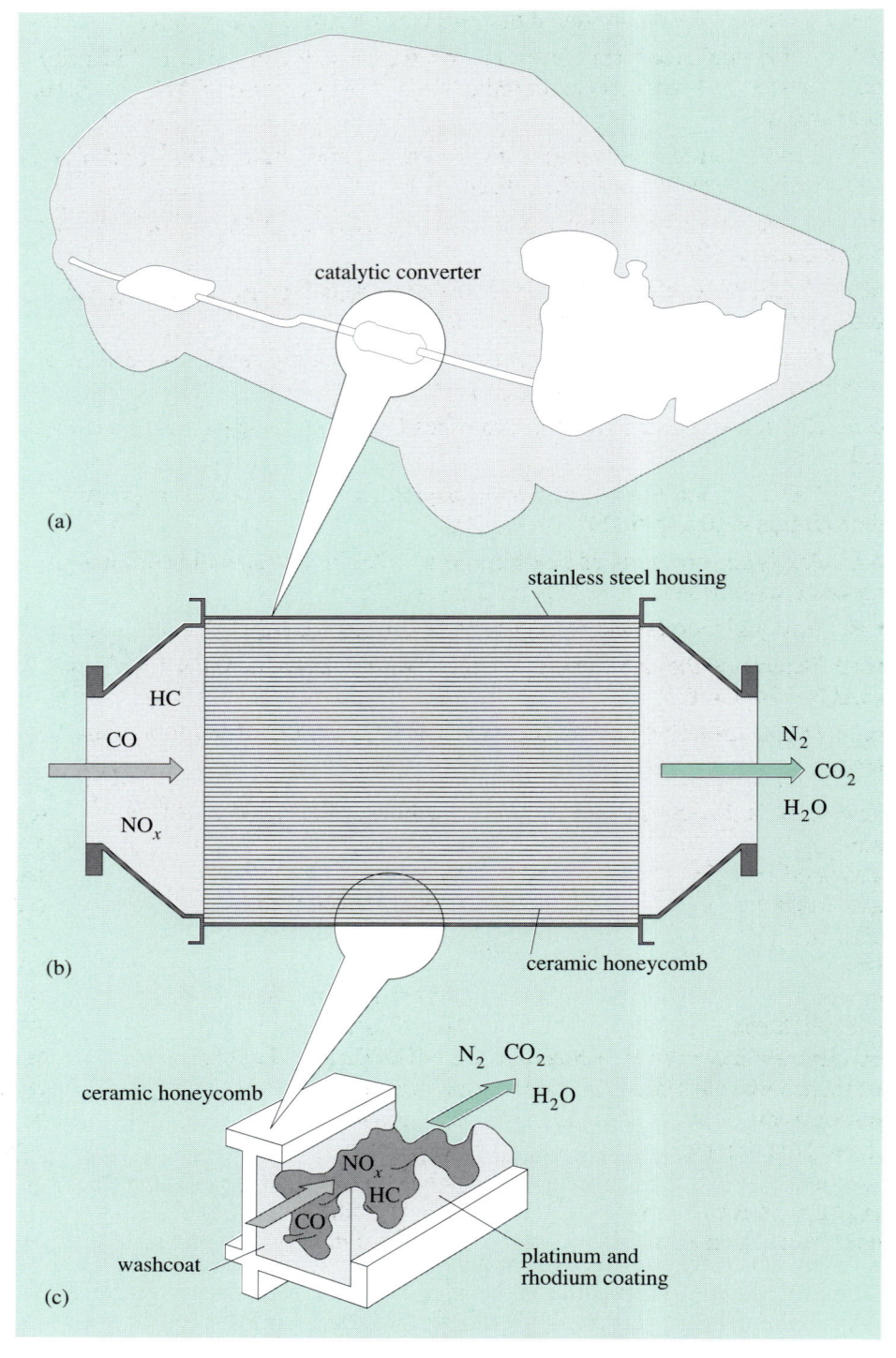

Figure 6.28 The catalytic convertor

Objectives

Having studied this Unit, you should now be able to do the following.

6.1 Write a balanced chemical equation for a specific reaction and identify how the states of reactants and products affect the outcome (SAQs 6.1, 6.10, 6.12 and 6.13).

6.2 Classify chemical reactions into oxidation, reduction, redox, hydrolysis, dissociation, condensation (SAQ 6.2).

6.3 Express reaction energies on a mass or molar basis (SAQs 6.3, 6.4, 6.5, 6.6, 6.7 and 6.8).

6.4 Calculate reaction enthalpies for specific reactions given appropriate information (SAQs 6.4, 6.5, 6.6, 6.7 and 6.8).

6.5 Assess the effect of the state of reactants or products on reactions in a quantitative thermodynamic way (SAQ 6.8).

6.6 Determine thermodynamic quantities from free energy diagrams (SAQ 6.9).

6.7 Calculate equilibrium constants or yields of reactions with appropriate data (SAQs 6.10 and 6.11).

6.8 Apply the concepts of equilibrium to solubility of specific compounds and acid/base behaviour (SAQs 6.12 and 6.13).

6.9 Apply kinetic concepts to chemical reactions (SAQs 6.14 and 6.15).

6.10 Determine the order and rate constant for simple chemical reactions (SAQs 6.14 and 6.15).

6.11 Define, describe or otherwise explain the meaning of the following terms:

acid dissociation constant	equilibrium reactions	reaction profiles
acids	ester	reaction rate
activated complex	exothermic	redox reaction
activation energy	first-order reaction	reducing agent
alkalis	Hess's law	reduction
bases	hydrolysis	reversible reactions
catalyst	ion product of water	solubility product
chemical equation	law of mass action	standard state
concentration equilibrium constant	Le Chatelier's principle	strong acid
condensation reactions	neutral solution	strong base
dissociation	oxidation	weak acid
Ellingham diagram	pH scale	weak base
endothermic	pseudo first–order	yield
enthalpy of formation	rate constant	
equilibrium constant	reaction enthalpy change	

Answers to exercises

EXERCISE 6.1 In the first point of the list, there is elemental iron, methyl chloride, magnesium oxide and hydrogen chloride; in the second, water, sodium sulphate, iron nitrate (it's actually ferrous, or iron (II) nitrate, but you're not expected to know that) and ammonium carbonate; and in the third, the cations of calcium, and iron (ferric or iron (III) in this case) and the oxide, sulphate and silicate anions.

EXERCISE 6.2 The thermal energy needed to raise 1 mole of liquid water from 298 K to 370 K is just

$75.4 \text{ J K}^{-1} \times (370 - 298) \text{ K} = 5429 \text{ J} \approx 5 \text{ kJ}$

So the *net* enthalpy change for the reaction is

$\Delta H = (-286 + 5) \text{ kJ mol}^{-1} = -281 \text{ kJ mol}^{-1}$

EXERCISE 6.3 In aqueous solution, sodium hydroxide dissociates according to the equation

$\text{NaOH(s)} \longrightarrow \text{Na}^+(\text{aq}) + \text{OH}^-$

Thus 0.1 mol l^{-1} of NaOH will produce 0.1 mol l^{-1} of OH^- ions. To find $[\text{H}^+]$ you use the ion product for water, K_w, given by:

$K_w = [\text{H}^+][\text{OH}^-] = 1 \times 10^{-14} \text{ mol}^2 \text{ l}^{-2}$

You've established that $[\text{OH}^-] = 0.1 \text{ mol l}^{-1}$, so

$[\text{H}^+] = \dfrac{1 \times 10^{-14} \text{ mol}^2 \text{ l}^{-2}}{[\text{OH}^-]}$

$= \dfrac{1 \times 10^{-14}}{0.1} \left(\dfrac{\text{mol}^2 \text{ l}^{-2}}{\text{mol l}^{-1}} \right)$

$= 1 \times 10^{-13} \text{ mol l}^{-1}$

Hence $[\text{OH}^-] = 0.1 \text{ mol l}^{-1}$ and $[\text{H}^+] = 1 \times 10^{-13} \text{ mol l}^{-1}$

Answers to self-assessment questions

SAQ 6.1
(a) $2Mg(s) + O_2(g) \longrightarrow 2MgO(s)$
(b) $2HNO_3(aq) + Ca(OH)_2(aq)$
$\longrightarrow Ca(NO_3)_2(aq) + 2H_2O$
(c) $4Fe(OH)_2(s) + O_2(g) + 2H_2O(l)$
$\longrightarrow 4Fe(OH)_3(s)$
(d) $Ag_2SO_4(aq) + 2e^-$
$\longrightarrow 2Ag(s) + SO_4^{2-}(aq)$

SAQ 6.2
(a) $2CuO(s) \xrightarrow{1720\,K} 2Cu(s) + O_2(g)$
is a reduction reaction with dissociation.
(b) $HNO_3 + H_2O \longrightarrow H_3O^+ + NO_3^-$
is a hydrolysis reaction with dissociation.
(c) $2Mg(s) + O_2(g) \longrightarrow 2MgO(s)$
is an oxidation reaction.
(d) $C_2H_5.COOH + CH_3.OH$
$\longrightarrow C_2H_5.COO.CH_3 + H_2O$
is an organic acid–base reaction, which also involves condensation.

SAQ 6.3
(a) The combustion equation is
$$H_2(g) + \tfrac{1}{2}O_2(g) \longrightarrow H_2O(l)$$
One mole of H_2 has a mass of 2 g. The combustion enthalpy of 2 g (0.002 kg) of hydrogen is −286 kJ. So the combustion enthalpy of 1 kg of hydrogen is
$$\Delta H = \left(-286 \times \frac{1}{0.002}\right) \text{kJ kg}^{-1}$$
$$= -143 \text{ MJ kg}^{-1}$$
(to 3 significant figures)

(b) The combustion equation is
$C_2H_4(g) + 3O_2(g)$
$\longrightarrow 2CO_2(g) + 2H_2O(g)$

The relative molecular mass of ethylene is $(2 \times 12) + (4 \times 1) = 28$.

So 1 mole of C_2H_4 has a mass of 28 g, or 0.028 kg. When 0.028 kg of C_2H_4 is burned, the combustion enthalpy is −1416 kJ. The combustion enthalpy of 1 kg of C_2H_4 is

$$\Delta H = \left(-1416 \times \frac{1}{0.028}\right) \text{kJ kg}^{-1}$$
$$= -50.6 \text{ MJ kg}^{-1}$$
(to 3 significant figures)

SAQ 6.4 The reactions given are:
$S(s) + O_2(g) \longrightarrow SO_2(g)$ (i)
$\Delta H^\ominus = -298 \text{ kJ (mol S)}^{-1}$

$SO_2(g) + \tfrac{1}{2}O_2(g) \longrightarrow SO_3(g)$ (ii)
$\Delta H^\ominus = -99 \text{ kJ (mol SO}_2)^{-1}$

You want $\Delta H^\ominus$ for the reaction
$S(s) + \tfrac{3}{2}O_2(g) \longrightarrow SO_3(g)$

Adding reactions (i) and (ii) gives:
$S(s) + O_2(g) + SO_2(g) + \tfrac{1}{2}O_2(g)$
$\longrightarrow SO_2(g) + SO_3(g)$

If you cancel the SO_2 terms on both sides of the equation and combine the oxygen terms you get
$S(s) + \tfrac{3}{2}O_2(g) = SO_3(g)$
with
$\Delta H^\ominus = (-298 - 99) \text{ kJ (mol S)}^{-1}$
$= -397 \text{ kJ (mol S)}^{-1}$

SAQ 6.5 The reactions given are:
$S(\text{monoclinic}) + O_2(g) \longrightarrow SO_2(g)$ (i)
$\Delta H^\ominus = -298.3 \text{ kJ (mol S)}^{-1}$

$S(\text{rhombic}) + O_2(g) \longrightarrow SO_2(g)$ (ii)
$\Delta H^\ominus = -298.0 \text{ kJ (mol S)}^{-1}$

You want $\Delta H^\ominus$ for the reaction
$S(\text{rhombic}) \longrightarrow S(\text{monoclinic})$

Reversing the monoclinic reaction (i) and adding it to the rhombic reaction (ii) gives:
$S(\text{rhombic}) + O_2(g) + SO_2(g)$
$\longrightarrow SO_2(g) + S(\text{monoclinic}) + O_2(g)$

Cancelling terms that appear on both sides of the equation gives:
$S(\text{rhombic}) \longrightarrow S(\text{monoclinic})$

$\Delta H^\ominus = (-298.0 + 298.3) \text{ kJ mol}^{-1}$
$= +0.3 \text{ kJ mol}^{-1}$

SAQ 6.6 The reaction is
$C_2H_2(g) + \tfrac{5}{2}O_2(g) \longrightarrow 2CO_2(g) + H_2O(l)$

The standard formation enthalpies from Tables 6.2 and 6.3 are:
$C_2H_2(g)$ $\Delta H_f^\ominus = +226.7 \text{ kJ mol}^{-1}$
$CO_2(g)$ $\Delta H_f^\ominus = -393.5 \text{ kJ mol}^{-1}$
$H_2O(l)$ $\Delta H_f^\ominus = -285.9 \text{ kJ mol}^{-1}$
and you know that
$O_2(g)$ $\Delta H_f^\ominus = 0$

Applying Equation (6.3)
$\Delta H_R^\ominus = \sum_{\text{products}} \Delta H_f^\ominus - \sum_{\text{reactants}} \Delta H_f^\ominus$
$= (2\Delta H_f^\ominus(CO_2) + \Delta H_f^\ominus(H_2O))$
$\quad - \left(\Delta H_f^\ominus(C_2H_2) + \tfrac{5}{2}\Delta H_f^\ominus(O_2)\right)$
$= (2 \times (-393.5) + (-285.9)) \text{ kJ mol}^{-1}$
$\quad - (+226.7) \text{ kJ mol}^{-1}$

$\Delta H_R^\ominus = -1300 \text{ kJ mol}^{-1}$
(to 4 significant figures)

Or, since this represents the combustion of one mole of C_2H_2,
$\Delta H_R^\ominus = -1300 \text{ kJ (mol C}_2H_2)^{-1}$

SAQ 6.7 The reaction is
$C_5H_{12}(g) \longrightarrow 2C_2H_4(g) + CH_4(g)$

From Table 6.3, the standard formation enthalpies are:
$C_5H_{12}(g)$ $\Delta H_f^\ominus = -134.5 \text{ kJ mol}^{-1}$
$C_2H_4(g)$ $\Delta H_f^\ominus = +52.3 \text{ kJ mol}^{-1}$
$CH_4(g)$ $\Delta H_f^\ominus = -74.8 \text{ kJ mol}^{-1}$

Applying Equation (6.3)
$\Delta H_R^\ominus = \sum_{\text{products}} \Delta H_f^\ominus - \sum_{\text{reactants}} \Delta H_f^\ominus$
$= (2\Delta H_f^\ominus(C_2H_4) + \Delta H_f^\ominus(CH_4))$
$\quad - (\Delta H_f^\ominus(C_5H_{12}))$
$= (2 \times (+52.3) + (-74.8)) \text{ kJ mol}^{-1}$
$\quad - (-134.5) \text{ kJ mol}^{-1}$
$= +164.3 \text{ kJ mol}^{-1}$
(to 4 significant figures)

$\Delta H^\ominus$ for this reaction is positive, so the reaction is endothermic.

SAQ 6.8

(a) The entropy of vaporization is given by the formula

$$\Delta S_v^\ominus = \frac{\Delta H_v^\ominus}{T_b}$$

The calculated values are shown in Table 6.15.

(b) The entropies of vaporization are on average a factor of ten greater than the entropies of fusion and there is relatively little difference between the various metals listed.

(c) The change in molecular disorder on vaporization of a liquid to a gas is very much larger than the melting of a solid to give a liquid. Both liquids and solids are condensed states of matter and have similar volumes, whereas the volumes of gases and liquids differ by a very large amount.

Table 6.15 Calculation of entropies of vaporization of selected metals

Material	$\frac{T_b}{K}$	$\frac{\Delta H_v^\ominus}{kJ\,mol^{-1}}$	$\frac{\Delta S_v^\ominus}{J\,mol^{-1}\,K^{-1}}$
aluminium	2740	285.4	104.2
copper	2868	305.8	106.6
iron	3008	355.3	118.1
tungsten	6200	775.3	125.0
mercury	630	59.2	94.0
titanium	3530	424.2	120.2
magnesium	1380	132.3	95.9
tin	2540	291.5	114.8

SAQ 6.9

(a) The temperature is the critical temperature where the line crosses the temperature axis. I made it about 715 K from the graph.

(b) The standard reaction enthalpy is obtained from the intercept on the $\Delta G^\ominus$ axis (see Figure 6.9). This is at about

$$\Delta G^\ominus = -155\,kJ\,(mol\,O_2)^{-1}$$

But the question asked for the value per mole of mercury. Since two moles of mercury react with each mole of oxygen, the value of $\Delta H^\ominus$ must be half the above,

or

$$\Delta H^\ominus = -77.5\,kJ\,(mol\,Hg)^{-1}$$

(c) The required standard entropy change is the slope of the graph between T_m and T_b (see Figure 6.9 again). I made it

$$\Delta S^\ominus = \frac{-180\,kJ\,(mol\,O_2)^{-1}}{840\,K}$$

$$= -214\,J\,(mol\,O_2)^{-1}\,K^{-1}$$

SAQ 6.10

(a) $C_2H_6(g) \rightleftharpoons C_2H_4(g) + H_2(g)$, hence

$$K_c = \frac{[C_2H_4][H_2]}{[C_2H_6]}$$

The units are:

$$\frac{(mol\,l^{-1})(mol\,l^{-1})}{(mol\,l^{-1})}$$

$$= \frac{(mol\,l^{-1})^2}{(mol\,l^{-1})}$$

$$= mol\,l^{-1}$$

(b) $2S(g) + 3O_2(g) \rightleftharpoons 2SO_3(g)$

$$K_c = \frac{[SO_3]^2}{[S]^2[O_2]^3}$$

The units are:

$$\frac{(mol\,l^{-1})^2}{(mol\,l^{-1})^2(mol\,l^{-1})^3}$$

$$= \frac{(mol\,l^{-1})^2}{(mol\,l^{-1})^5}$$

$$= (mol\,l^{-1})^{-3}$$

SAQ 6.11 The balanced equation for the reaction is:

$$H_2(g) + I_2(g) \longrightarrow 2HI(g)$$

Initially

$[H_2] = [I_2] = 1\,mol\,l^{-1}$

$[HI] = 0$

If, in reaching equilibrium, x moles of hydrogen combine with iodine to form HI, then, from the equation, x moles of iodine must also have reacted and $2x$ moles of HI must have been produced. Hence, at equilibrium

$[H_2] = [I_2] = (1 - x)\,mol\,l^{-1}$

$[HI] = 2x\,mol\,l^{-1}$

The equilibrium constant for this reaction is defined by

$$K_c = \frac{[HI]^2}{[H_2][I_2]} \frac{(mol\,l^{-1})^2}{(mol\,l^{-1})(mol\,l^{-1})}$$

$$= \frac{(2x)^2}{(1-x)(1-x)}$$

$$= \frac{(2x)^2}{(1-x)^2}$$

But under the conditions of the reaction $K_c = 0.040$, so

$$\frac{(2x)^2}{(1-x)^2} = 0.040$$

Take square roots of both sides:

$$\frac{2x}{1-x} = 0.20$$

or

$2x = 0.20(1 - x)$

$= 0.20 - 0.2x$

$2.2x = 0.20$

$$x = \frac{0.20}{2.2}$$

$= 0.091$

Hence the equilibrium yield is only 9.1% of that for the complete reaction.

SAQ 6.12

(a)

$$KCl(s) \underset{}{\overset{H_2O}{\rightleftharpoons}} K^+(aq) + Cl^-(aq)$$

$K_{sp} = [K^+][Cl^-]$

Units of K_{sp} are

$(mol\,l^{-1})(mol\,l^{-1}) = (mol\,l^{-1})^2$ or $mol^2\,l^{-2}$

(b)

$$BaSO_4(s) \underset{}{\overset{H_2O}{\rightleftharpoons}} Ba^{2+}(aq) + SO_4^{2-}(aq)$$

$K_{sp} = [Ba^{2+}][SO_4^{2-}]$

Units of K_{sp} are

$(mol\,l^{-1})(mol\,l^{-1}) = (mol\,l^{-1})^2$ or $mol^2\,l^{-2}$

(c)

$$Ca(OH)_2(s) \underset{}{\overset{H_2O}{\rightleftharpoons}} Ca^{2+}(aq) + 2OH^-(aq)$$

$K_{sp} = [Ca^{2+}][OH^-]^2$

Units of K_{sp} are

$(mol\,l^{-1})(mol\,l^{-1})^2 = (mol\,l^{-1})^3$ or $mol^3\,l^{-3}$

SAQ 6.13 Calcium sulphate dissociates according to the reaction

$$CaSO_4(s) \xrightleftharpoons{H_2O} Ca^{2+}(aq) + SO_4^{2-}(aq)$$

If the solubility of calcium sulphate is x mol l^{-1}, then x mol l^{-1} of CaSO$_4$ will produce x mol l^{-1} of Ca^{2+} and x mol l^{-1} of SO$_4^{2-}$.

Solubility product K_{sp} is given by

$$K_{sp} = [Ca^{2+}][SO_4^{2-}] \,(\text{mol l}^{-1})^2$$
$$= x^2 \text{ mol}^2 \text{ l}^{-2}$$

But $K_{sp} = 2.0 \times 10^{-5}$ mol^2 l^{-2}, so

$$x^2 = 2.0 \times 10^{-5} \text{ mol}^2 \text{ l}^{-2}$$
$$x = 4.5 \times 10^{-3} \text{ mol l}^{-1}$$

SAQ 6.14 The fact that in this example concentration is measured as a pressure should not deter you. You should proceed in exactly the same way as the example in the text. The first stage is to obtain values of $\ln[N_2O_5]$ so that you can plot the graph of $\ln[N_2O_5]$ versus time. This is shown in Figure 6.29. Note that it helps to choose the ranges of your axes to match the data; if you take the axes in full from the origin, the graph will appear very small.

The fact that the graph is linear confirms that the reaction is first order. The slope of the graph is easily determined because the straight line passes through all of the plotted points, so you can work from the highest and lowest data values.

$$\text{Slope} = \frac{\ln 33.5 - \ln 8.4}{1200 - 12\,000} \text{ s}^{-1}$$
$$= \frac{1.383}{-10\,800} \text{ s}^{-1}$$
$$= -1.3 \times 10^{-4} \text{ s}^{-1}$$

(to 2 significant figures)

So $k = 1.3 \times 10^{-4}$ s^{-1} (note, slope $= -k$).

SAQ 6.15 Since $kt = \ln[A]_0 - \ln[A]$ for a first-order reaction, you can simply insert the methyl acetate concentration at any arbitrary interval of time using the data of Table 6.13 (a procedure equivalent to measuring the slope of the line of Figure 6.21).

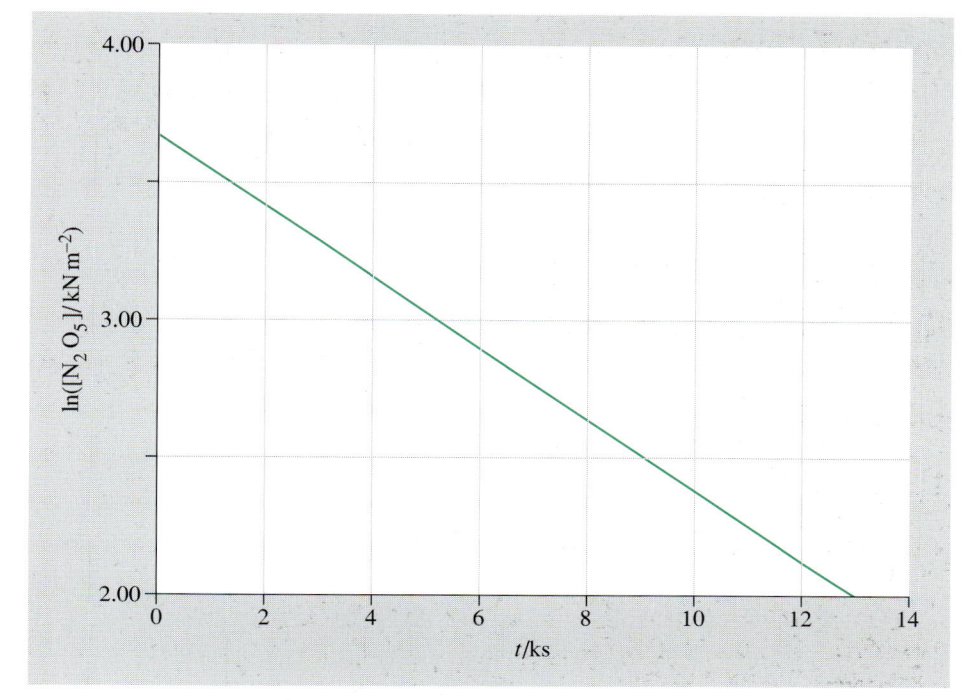

Figure 6.29 Plot of $\ln[N_2O_5]$ versus t for SAQ 6.14

$$k \times 1200 = \ln\left(\frac{0.308}{0.288}\right)$$

hence

$$k = \frac{0.0671}{1200} = 5.60 \times 10^{-5} \text{ s}^{-1}$$

(a) [CH$_3$COO.CH$_3$] at any time can be calculated in a similar way knowing k. At 4 hours,

$$t = 60 \times 60 \times 4 = 14\,400 \text{ s}$$

so

$$\ln \frac{0.308}{[CH_3COO.CH_3]}$$
$$= 5.60 \times 10^{-5} \times 14\,4000$$
$$= 0.806$$

Then

$$\frac{0.308}{[CH_3COO.CH_3]}$$
$$= \text{antiln } 0.806$$
$$= 2.240$$

Hence

$$[CH_3COO.CH_3]$$
$$= \frac{0.308}{2.240} \text{ mol l}^{-1}$$
$$= 0.138 \text{ mol l}^{-1}$$

(to 3 significant figures)

So the concentration of methyl acetate after 4 hours is 0.138 moles per litre.

(b) Since the reaction equation shows that one mole of methyl acetate produces one mole of acetic acid on hydrolysis, the concentration of acetic acid after four hours must be

$$([CH_3COO.CH_3]_0 - 0.138) \text{ mol l}^{-1}$$

Hence the concentration of acetic acid after four hours hydrolysis is just

$$0.308 - 0.138 = 0.170 \text{ mol l}^{-1}$$

Acknowledgements

Grateful acknowledgement is made to the following for permission to reproduce illustrations used in this book.

AV1/2 Frame 1 (left) © Royal Society of Chemistry. AV1/2 Frame 1 (right) © Image Select.

Figure 6.1(a),© Du Pont. Figure 6.1(b) © J. Allan Cash. Figure 6.2(a), © British Steel, Teeside Works. Figure 6.11, © Mansell Collection. Figure 6.13, © Trustees of the Science Museum. Figures 6.14 and 6.23 adapted from E. N. Ramsden, *A-Level Chemistry*, 2nd edn, Stanley Thornes (Publishers) Ltd 1990.

UNIT 6   CHEMICAL REACTIONS

THE PERIODIC TABLE

1a	2a	3b	4b	5b	6b	7b	8		
1 H $1s^1$									
3 Li $-2s^1$	4 Be $-2s^2$								
11 Na $-3s^1$	12 Mg $-3s^2$								
19 K $-4s^1$	20 Ca $-4s^2$	21 Sc $-4s^2 3d^1$	22 Ti $-4s^2 3d^2$	23 V $-4s^2 3d^3$	24 Cr $-4s^1 3d^5$	25 Mn $-4s^2 3d^5$	26 Fe $-4s^2 3d^6$	27 Co $-4s^2 3d^7$	
37 Rb $-5s^1$	38 Sr $-5s^2$	39 Y $-5s^2 4d^1$	40 Zr $-5s^2 4d^2$	41 Nb $-5s^1 4d^4$	42 Mo $-5s^1 4d^5$	43 Tc $-5s^1 4d^6$	44 Ru $-5s^1 4d^7$	45 Rh $-5s^1 4d^8$	
55 Cs $-6s^1$	56 Ba $-6s^2$	57 to 71 *	72 Hf $-6s^2 4f^{14} 5d^2$	73 Ta $-6s^2 4f^{14} 5d^3$	74 W $-6s^2 4f^{14} 5d^4$	75 Re $-6s^2 4f^{14} 5d^5$	76 Os $-6s^2 4f^{14} 5d^6$	77 Ir $-6s^0 4f^{14} 5d^9$	
87 Fr $-7s^1$	88 Ra $-7s^2$	89 to 92 †							

	57 La	58 Ce	59 Pr	60 Nd	61 Pm	62 Sm	63 Eu
* Lanthanides							

	89 Ac	90 Th	91 Pa	92 U
† Actinides				

	1b	2b	3a	4a	5a	6a	7a	0
METALS			SEMI-METALS	NON-METALS				2 He $1s^2$
			5 B $-2s^22p^1$	6 C $-2s^22p^2$	7 N $-2s^22p^3$	8 O $-2s^22p^4$	9 F $-2s^22p^5$	10 Ne $-2s^22p^6$
			13 Al $-3s^23p^1$	14 Si $-3s^23p^2$	15 P $-3s^23p^3$	16 S $-3s^23p^4$	17 Cl $-3s^23p^5$	18 Ar $-3s^23p^6$
28 Ni $-4s^23d^8$	29 Cu $-4s^13d^{10}$	30 Zn $-4s^23d^{10}$	31 Ga $-3d^{10}4p^1$	32 Ge $-3d^{10}4p^2$	33 As $-3d^{10}4p^3$	34 Se $-3d^{10}4p^4$	35 Br $-3d^{10}4p^5$	36 Kr $-4s^23d^{10}4p^6$
46 Pd $-5s^04d^{10}$	47 Ag $-5s^14d^{10}$	48 Cd $-5s^24d^{10}$	49 In $-4d^{10}5p^1$	50 Sn $-4d^{10}5p^2$	51 Sb $-4d^{10}5p^3$	52 Te $-4d^{10}5p^4$	53 I $-4d^{10}5p^5$	54 Xe $-5s^24d^{10}5p^6$
78 Pt $-6s^14f^{14}5d^9$	79 Au $-6s^14f^{14}5d^{10}$	80 Hg $-6s^24f^{14}5d^{10}$	81 Tl $-6p^1$	82 Pb $-6p^2$	83 Bi $-6p^3$	84 Po $-6p^4$	85 At $-6p^5$	86 Rn $-6s^26p^6$
							NON-METALS	

METALS

64 Gd	65 Tb	66 Dy	67 Ho	68 Er	69 Tm	70 Yb	71 Lu

SCIENCE FOR MATERIALS

The Greek alphabet

Lower case	Upper case	Name	Used for (lower case unless specified)
α	A	**alpha**	thermal expansion coefficient, an angle
β	B	**beta**	an angle
γ	Γ	**gamma**	shear strain, an angle
δ	Δ	**delta**	small increment (δx), 'change in' (Δy)
ε	E	**epsilon**	tensile strain, permittivity
ζ	Z	**zeta**	—
η	H	**eta**	coefficient of viscosity
θ	Θ	**theta**	an angle
ι	I	**iota**	—
κ	K	**kappa**	thermal conductivity
λ	Λ	**lambda**	wavelength
μ, μ	M	**mu**	coefficient of friction, micro- (μ)
ν	N	**nu**	Poisson's ratio, frequency
ξ	Ξ	**xi**	—
o	O	**omicron**	—
π	Π	**pi**	basic mathematical constant
ρ	P	**rho**	density, electrical resistivity (ρ_e)
σ	Σ	**sigma**	tensile stress, electrical conductivity, sum (Σx)
τ	T	**tau**	shear stress, time constant
υ	Y	**upsilon**	—
ϕ	Φ	**phi**	an angle
χ	X	**chi**	—
ψ	Ψ	**psi**	—
ω	Ω	**omega**	angular frequency, ohm (Ω)